THE SIMPLICITY CYCLE

ALSO BY DAN WARD

F.I.R.E.: How Fast, Inexpensive, Restrained, and Elegant Methods Ignite Innovation

THE SIMPLICITY CYCLE

A Field Guide to Making Things Better Without Making Them Worse

DAN WARD

HARPER
BUSINESS

An Imprint of HarperCollins*Publishers*

HarperCollins books may be purchased for educational, business, or sales promotional use. For information, please e-mail the Special Markets Department at SPsales@harpercollins.com.

xkcd comic reprinted by permission.

Excerpt from Cliff Crego's "On Complexity, Simplicity and Human Design" is reprinted with permission.

Quote from Neil Mix's blog post reprinted with permission.

All other images and diagrams are by the author.

FIRST EDITION

Designed by Renato Stanisic

Library of Congress Cataloging-in-Publication Data has been applied for.

ISBN: 978-0-06-230197-0

15 16 17 18 19   OV/RRD   10 9 8 7 6 5 4 3 2 1

Men rush toward complexity; but they yearn for simplicity.

—G. K. CHESTERTON

CONTENTS

CONTENTS

LIST OF FIGURES

FOREWORD
By Don Norman

This is a simple book about a complex topic, but don't be fooled. Beneath the simplicity lies a deep and profound message. Complexity is often necessary, but unnecessary complexity complicates our lives. How can we strike the proper balance? Ah. Start with this delightful book.

I first encountered Dan Ward in 2008 when we corresponded about an early edition of this book. I was working on my own book about complexity and it soon became clear that we shared similar viewpoints. An officer in the U.S. Air Force, Dan was working to streamline the way that complex projects are developed within the military. In 2012 he wrote that "I was just offered a position in D.C. They want me to help lead an initiative to implement my FIST [Fast, Inexpensive, Simple, Tiny] concept across the U.S. government. . . ." Neat. This meant that he didn't just have a book, he had a method. Moreover, he was being given an opportunity to implement the plan in major projects.

FIST, his procedure for developing projects quickly and efficiently, requires that things be less physically complex, less cognitively confusing, and less complicated. In 2014, FIST was renamed F.I.R.E. (for Fast, Inexpensive, Restrained, and Elegant) and published as a book by HarperCollins. This book provides a deeper look at the simple/elegant portion of his framework.

Naturally, the two books support one another. *F.I.R.E.* offers lessons highly relevant to *The Simplicity Cycle*. Large projects all tend to fail. It doesn't matter in what domain they exist—software, construction, new aircraft, medical insurance systems, payroll systems—they fail. Ward offers a simple solution: don't do them. With the time and money allocated for one large project, do numerous small ones. Do them Fast and Inexpensive, with Restraint and Elegance: F.I.R.E.

It's a well-known principle, but it goes against the nature of organizations that wish to solve all their problems with one project. In consumer markets, it encourages the disease I call featuritis. In industry, it's bloat. The military calls it requirements creep. What's the alternative? F.I.R.E. How do we avoid unneeded complexity and manage to maintain simplicity? That's the focus of this book.

In *The Simplicity Cycle*, Ward examines the normal trajectory of the design of systems. Consider how good a system might be, perhaps with some measure of "goodness." To increase goodness, complexity must be increased, because zero complexity generally means zero goodness. At some point however, the complexity starts getting in the way, perhaps by making the system far too difficult to design, or to construct, or to manage. Perhaps it is now too difficult to comprehend. Whatever the reason, once the design goes past a certain point, increasing complexity begins to make things worse. But it is often necessary to reach the point of overcomplexity in order to get the balance right. As Ward puts it:

> Patience and diligence are keys to avoiding premature optimization. First, we need to gain the necessary tools, talents, pieces, parts, and components . . . and only then can we apply them in the appropriate degree and trim out the extraneous.
>
> But simplifying too soon is just as bad as complexifying for too long.

Ward suggests several ways of reducing unneeded complexity. One is simply to begin removing components and see how well the system functions without them. If it still functions well, those components were unnecessary. Keep doing this until as much as possible has been removed while still yielding acceptable goodness. Does this sound too simple? Too obvious? Well, it is still resisted. Ward describes the result this way:

> Resistance to simplification is based on the belief that every additional feature, part, and function represents an improvement. It also assumes that the accumulated additions made things better from an overall system perspective.
>
> Such assertions are misguided and maybe a little arrogant.
>
> Why "arrogant"? Because they assume everything we ever added was a good idea. Why "maybe"? Because those additions may indeed have been good. But getting rid of them might be even better.

A second method is through restructuring. "A cube is less complex than a collection of squares," he points out, because "it is one object, not six. It is also 'more good' because we can do more with a 3-D object than a 2-D object." This is similar to the transformation of airline cockpits from many separate, physical displays into a smaller number of well-integrated visual displays—what is today called "the glass cockpit." It wasn't just a matter of combining things; it required rethinking them, reconceptualizing them. The result was that complexity went down while goodness went up. Similarly, taking six squares and using them to form a cube reconceptualizes it as a new object, where we no longer think of it as having six separate parts but rather as a single, integrated whole. Complexity goes down, goodness goes up.

The two books, *F.I.R.E.* and *The Simplicity Cycle*, can each be read on its own, but for people involved in the design and implementation of complex projects, they form a powerful pair. For anyone wanting to embrace the mantle of simplicity, this book, *The Simplicity Cycle*, is essential.

Making something simple is difficult. Simplicity is actually quite complex.

Don Norman
Silicon Valley, California
Author of *Living with Complexity* and *The Design of Everyday Things*, Revised and Expanded

CHAPTER 1

Soda, Swordsmen, and Road Maps

Complexity is a two-liter bottle of soda.

Doled out in reasonable quantities and at appropriate times, it's not bad. In fact, it can be pleasant as an occasional accompaniment to a balanced meal or as a refreshing treat on a hot day. Unfortunately, we're guzzling gallons of the stuff every day and it's killing us.

The fact that humans have a taste for sweetness is neither a genetic flaw nor a psychological disorder. It's a survival trait, passed down from our ancestors who needed to find high concentrations of calories in order to stay alive and thus developed an affinity for sugar. This preference became a problem in the modern era, where we have easy access to unlimited quantities of cheap, industrial-strength concentrations of carbonated high-fructose corn syrup, packaged in attractively colored cans and bottles. Instead of survival, we ended up with obesity and diabetes.

Similarly, humans gravitate toward complexity, in our technologies and religions, our laws and relationships, because simplicity is so often inadequate to our needs. We require a certain degree of complexity in our lives, just as we require a certain number of calories each day. Accordingly, we add layers, gizmos, features, functions, connections, and rules to the things we create in an attempt to make them more exciting, more effective, or otherwise better. This preference, too, becomes a problem when it spirals out of control and produces industrial-strength concentrations of complexity that surpass our needs by multiple orders of magnitude. Press 1 if you agree. Press 2 for a list of other options. Press 3 to return to the main menu. Please note, the options have changed.

As with using fancified sugar water to satisfy a sweet tooth, our otherwise healthy predisposition toward complexity goes awry when it is manipulated by marketers or thoughtlessly indulged by designers and engineers who fail to foresee the unproductive consequences of their actions.

The ill-advised ubercomplexity we so often encounter is more than a time-wasting nuisance. In some cases, it has life-and-death consequences. In the medical field unnecessary complexity leads to wasted resources, ballooning costs, delayed treatments, and entirely avoidable complications (both medical and procedural). As Dr. Atul Gawande explains in his book *The Checklist Manifesto*, "[T]he source of our greatest difficulties and stresses in medicine . . . is the complexity that science has dropped upon us and the enormous strains we are encountering in making good on its promise." He goes on to observe that "defeat under conditions of complexity occurs far more often despite great effort rather than from a lack of it."

The problem isn't that we aren't straining enough. It's that even our best efforts cannot overcome the weight of complexity. The solution may not require pushing harder. Instead, it may require rethinking our approach to complexity in the first place.

Comparable problems can be found in education, where the formalized complexity of educational policies actively interferes with scholarship and makes it harder for Johnny to learn how to read, while overengineered (and underdesigned) technical tools provide teachers and students with shiny objects that distract more than they help. The result is an expensive rush toward less learning, not more. In contrast, the entirely unofficial Kahn Academy provides an online repository of free educational videos, powerfully simple resources used by 10 million students each month (including my kids!).

The same thing happens in fields as diverse as law, engineering, energy production, and social services. Time and again, excessively complicated tools reduce our aptitude, and well-intentioned increases in complexity make things predictably worse. Even when the consequences are not medically dire, complexity reduces transparency and makes it difficult to see what is really going on. When effective alternatives like the Kahn Academy exist, they tend to be less industrial and more organic.

What can be done about it? Banning complexity outright would be both unwise and impossible. Frankly, we'd have an easier time forbidding the sale of sixty-four-ounce sodas in New York City. But while most of us cannot directly simplify things like the tax code or the iTunes End User License Agreement, we do not need to resign ourselves to the confusion, frustration, and waste that come from overcomplexification. There are things we can do to improve our lives and the lives of those around us.

This is particularly true when we design something, whether the thing is as ephemeral as an email or as enduring as a skyscraper. We can refuse to accept high levels of complexity as inevitable and refuse to view these levels as desirable. But even as consumers we have an opportunity to make things better by voting with our funds and purchasing simpler, more elegant alternatives whenever possible. Such alternatives sometimes cost less to purchase but almost always cost less to own, because they perform more reliably and effectively than the more complex options. Plus, these simpler alternatives make us happier, and that counts for something.

In *The Book of Five Rings*, the great sixteenth-century Japanese swordsman Miyamoto Musashi wrote, "From one thing, know ten thousand things." There is powerful simplicity in the idea that one truth can illuminate ten thousand other truths, and I'm sure our *ronin* friend is correct. The challenge is to identify and express a single foundational concept upon which "ten thousand things" rely.

In the spirit of Musashi I would like to offer one such truth for your consideration, a principle that provides a touchstone for subsequent decision-making and design approaches. The great truth that will equip us with wisdom for our journey and help us find our way—or ten thousand ways—through the labyrinth of complexity can be expressed in five words: simplicity is not the point.

Yes, in a world of ever-increasing complications, it is tempting to hold up simplicity as a cardinal virtue, a universally desirable attribute for all designed things. This perspective is understandable and excusable, and it often produces good results in the short term. However, it's also wrongheaded by about 90 degrees and contains the seeds of much subsequent failure.

Simplicity is great and important, to be sure, but let me say it again: simplicity is not the point. What is the point? In a word, *goodness*. Whether we are designing software or spacecraft, presentations or pizzas, the objective is to create something "good." Simplicity matters because it affects goodness, but it turns out that the relationship between simplicity and goodness does not follow a straight line. This means an increase in one does not always correspond with an increase in the other. Sometimes making things simpler is indeed an improvement. Sometimes not. Life is tricky that way.

Ultimately, it does not matter how simple or complex something is. The only question is whether the thing is any good. For example, my favorite homemade bread recipe has just four ingredients—water, flour, salt, and yeast. It's yummy and remarkably easy to make, and there is nothing I could add to the dough that would make it better. In its simplicity, it is just about perfect.

Having said that, I also love a good multigrain loaf, with five, seven, or even twelve different grains, plus a scattering of sunflower seeds and other tasty ingredients. This more complex bread is also yummy, although instead of baking it myself, I buy it from a shop.

When I eat these two types of bread, thoughts of simplicity and complexity are the last thing on my mind. I just enjoy them because they are both so very good. The simplicity or complexity contribute to the quality, but the *quality* is the appealing factor. Again, simplicity is not the point. Goodness is the point.

Who gets to define goodness? The customer, of course. Sure, the engineers, inventors, chefs, and designers all have a say, and their informed opinion matters greatly. Same goes for the business leaders and visionaries responsible for guiding the effort. But the customer has the last word on whether or not the design is good, and that is an important truth to keep in mind.

So goodness matters more than simplicity, but the two attributes are connected in important ways. As we seek to make things "more gooder," it helps to understand how goodness and simplicity are related. Specifically, it helps to recognize a few critical pivot points where activities that previously drove improvements begin to instead make things worse, where complexity becomes counterproductive, or where simplicity is inadequate. That's where this book comes in.

The pages that follow aim to identify some of these points, providing a road map that highlights the good paths and identifies the dead ends we might encounter on a journey of design. Think of it as an atlas, showing a wide swath of geography from various perspectives and explaining that a left turn in Albuquerque will take us toward a certain place, while a right turn will take us somewhere else.

Which turn should you take? That depends on where you're coming from and where you want to go. An atlas can point out nice places to visit and identify the fastest or most scenic way to get there, but it can't pick your destination for you, nor can it dictate which route you should take. Such decisions rightly belong to the traveler. That's part of the job, and it's also a big part of the fun.

As you read, keep in mind that the map is not the territory, and studying a map is no substitute for an actual expedition. The only way to *really* know what is out there is to go see for yourself. If you choose to set out on some sort of design adventure, don't be surprised when you encounter bumps in the road, unexpected twists, and various landscape features that escaped the attention of your humble cartographer. But as you head out into unknown territory—and design is always unknown—it may be helpful to bring a map along and consult it from time to time. It may also be wise to spend some time with the map *before* the journey begins, to get an idea of how long the trip might be, what gear to pack, and what to expect along the way.

How we use a map depends on who we are, where we are trying to go, and how we plan to get there. Thus, a biker, a hiker, and a truck driver might all use the same map but in vastly different ways, paying attention to different features and arriving at different destinations. Those traveling by foot may pay more attention to the space between roads, looking for dotted lines that indicate casual trails through wooded areas. Meanwhile, the people who rely on two-wheeled, self-powered vehicles to travel short distances would be wise to avoid the four-lane interstates and instead stick to the smaller, slower byways. But for the people who drive a big rig on eighteen wheels, those major highways are just about perfect.

Even people with a common transportation mode may approach the same map with diverse interests, because when you're driving in a street race your needs and objectives are not the same as when it's your turn to drive the car pool. Or maybe I'm wrong about that—I don't know much about your car-pool arrangements.

If this book is a map, who is the map for? It is for anyone who designs things, and that is pretty much all of us. As Henry Petroski explained in his book *Small Things Considered*, "We think, therefore we design. Indeed, there is barely anything we do, much less use, that does not have a design component to it."

Thus, this book is also for anyone who buys, consumes, or otherwise uses things, which is also all of us. Petroski argues that simply using a product often involves a number of design decisions. So, if you're interested in reducing confusion, frustration, and waste—yours or someone else's—whether as a designer or a consumer, dealing with hardware or software, services, or processes, then this map is for you.

The specifics of how you might use the map will vary depending on your needs. Coders, engineers, and other technologists may find it useful to guide technical design decisions and to shape the framework of their system architectures. Writers may discover that the process of literary creation is subject to many of the same complexity-related pitfalls and opportunities as a technical design project and thus may find familiar territory in the map. So, too, with people who prepare food, give presentations, or make plans.

If your business is business, you almost certainly face situations where complexity threatens to overwhelm value, where you must make decisions about a process, a customer, or an organizational structure that strike a balance between too much and not enough, between needs and wants and do-not-wants. These decisions require an understanding of how complexity affects quality and performance. This book is for you.

And finally, a word about leadership. A leader's job is in some sense to see the future, to cast a vision of what could be. While looking ahead to tomorrow's challenges and opportunities, leaders must simultaneously guide their teams along the path today, toward a finish line that may be invisible to most others. A good map is indispensable in such a situation, both to help the leader make good decisions about how to proceed and also to help discuss the plan with the rest of the team.

"See, we are *here*," the leader might say to her followers as she points to a spot on the map. "In order to get over *there*, we could go *this* way or *that* way." A map provides context for her words and gestures, making it easier for followers to engage with, understand, and embrace the vision.

All of this begs the question of whether anyone really needs a map. Can't we solve the problem of complexity by following the KISS principle of "Keep it simple, stupid"? What could be easier than that, right? And yet, maintaining simplicity is not only difficult, it is often ill-advised. In virtually any context—business planning, software development, pizza making—the immature simplicity associated with first drafts, early prototypes, and initial versions tends to be unsatisfyingly empty. Such partial solutions are necessary starting points but inadequate as final products because they lack the essential qualities found in more mature versions. Keeping things simple impedes progress, which is why we add things to our designs. We make them more complicated in order to make them better. It is only when we go too far that an epic level of complexity makes the final product unusable.

And this is where things can get a little tricky. Just because there is such a thing as too simple and such a thing as too complex does not mean the best solution sits in a mythical sweet spot between simplistic and complicated, as if the territory in question could be represented by a single straight line. Reality is more complicated than that, and straight lines seldom make for interesting journeys. There are bends in the road we need to be aware of. If we overlook them we'll either end up following the road to a bad part of town or jumping off the tracks entirely. Thus, the map.

I encountered one of these bends in the road in 2002, as I pitched my latest project to the U.S. Navy. Before I launched into my demonstration, the lieutenant commander across the table held up her hand and made a comment that changed my life.

"Wait. Before you begin, I want to say something. I don't care how good this thing is. If it isn't easy to use, I don't want it."

I sat back in my chair for a moment and thought about what I had just heard. She doesn't care how good the project is? What a remarkable statement! All the cool features and functions in the world wouldn't be interesting to her team, would not qualify as desirable qualities, unless the overall system was simple. She was saying complexity trumps goodness, a sentiment that resonated with me as much as it challenged, surprised, and confused me.*

Neither of us knew it at the time, but her comment led me to embark on a long investigation into the relationship between complexity and goodness in engineering and design. How do complexity and goodness really affect each other? Is simplicity the point after all? My Navy friend certainly seemed to think so,

---

In case you're wondering, the project I demonstrated that day was easy to use . . . and she liked it.

but was she correct? Did I even understand her comment? What did it really mean? Clearly, I had more research to do.

Shortly after that meeting, a number of other things came together, converging into the ideas that eventually became the Simplicity Cycle. I encountered the works of poet Cliff Crego, which we'll take a look at shortly. I had some hallway conversations with engineers, and conference-room discussions with senior managers. At some point it occurred to me that simplicity and ease of use aren't necessarily the same thing.

Through it all, I felt I was on the verge of understanding something about design, complexity, and utility. I made a few attempts to represent the idea visually and sketched out a series of little diagrams, none of which looked quite right. Something was still missing.

The journey continued long after that conversation. I talked with more people, read more books, and experimented with different design approaches, and things began to fall into place. I learned things. I unlearned things. I gathered new pieces of information. I set pieces aside and put others together. I can't point to a single breakthrough moment where the lightbulb came on, but I eventually sketched out the diagram shown in Figure 1. As I reflected on the drawing, I realized it not only described the process of designing something—an airplane, an outfit, a computer program, a presentation, a book; it also described the intellectual path I'd traveled in producing the diagram itself.

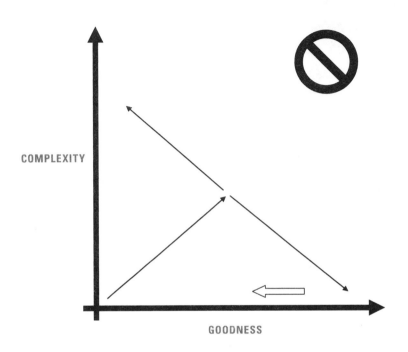

FIGURE 1: THE SIMPLICITY CYCLE

Don't worry if this figure doesn't make sense yet. It's not supposed to (yet). That's part of the deal. In the pages that follow we will step through each piece, then fit the pieces together. It'll become clearer once we define some terms and introduce some labels. We'll do that now.

CHAPTER 2

The Journey Begins

Like a movie star who needs no introduction, complexity is a central concept that feels familiar to us all. We've seen it often enough that we feel as if we know it. But also like a movie star, complexity contains layers that are hidden beneath the surface and that most of us are unaware of. Upon closer inspection, we discover that complexity is, well, more complex than it appears. In fact, *complexity* turns out to be one of those words with many definitions, some more involved and convoluted than others. If we're not careful, that can get confusing.

When a consumer says something is complex, it's usually not a compliment. But when connoisseurs use the word *complexity* to describe the flavor attributes of things like coffee, wine, or chocolate, they almost always mean it in a positive sense and place a premium on the product's depth and character. This type of complexity is regarded by experts as an essential element of excellence . . . except for when it isn't.

Too much complexity in our comestibles devolves into complicatedness, leading critics to complain that the flavors are aggressive, messy, and off-putting. Wine seller and blogger Joe Appel explains: "Complicated wines . . . don't arc toward harmony the way great, complex wines do. Complicated wines are more unsettled. . . . Not everything resolves; the mechanics are clunkier." His comments could just as easily be about software or consumer electronics rather than fermented fruit juice. But despite his critique that complicated wines do not pair well with any particular food and usually trigger an urge to run screaming from the room, he goes on to admit that "every once in a while, what you despise about something becomes a reason to love it." Yes, a difficult, confusing, jumbled wine can occasionally be exactly what the doctor ordered. Clearly, our relationship with complexity is complicated.

Physicists and mathematicians, meanwhile, use the word with great precision but without passing judgment—the mathematical complexity of the physical world is neither good nor bad; it is simply observed, measured, and documented. Some academics find it useful to draw a distinction between *detail* complexity and *dynamic* complexity, while others talk about things like Kolmogorov complexity and thermodynamic depth. Don't worry—we won't need those concepts here.

The elaborate, specialized, and overlapping definitions used by different groups are no doubt useful and necessary in their own contexts, but they are not particularly useful for the discussion in this book. Instead, we'll use a deliberately simple definition of the word, based on its general, nonscientific usage. The definition goes like this:

Complexity: Consisting of interconnected parts

Lots of interconnected parts equal a high degree of complexity—think of the world economy and the interplay between nations, companies, technology, and even the weather. For that matter, just think of weather itself, which is the product of interactions between such a bewilderingly diverse set of components (geography, solar activity, human activity, air, water, etc.) that we can barely get an accurate forecast for next weekend, let alone a prediction of which days next month will be rainy.

In contrast, few interconnected parts equal a low degree of complexity. Consider a staple holding several pieces of paper together. Their interaction is stable and unambiguous, easily described and understood. With me so far? Great. But here's the bad news: this approach to defining complexity isn't as simple as it appears.

The problem has to do with the word *lots*. That's a very subjective word. For better or worse, there is no absolute scale we can use to rigorously define *lots*, as if any number of parts larger than X constitutes a lot.

Even though *consisting of interconnected parts* is a simple definition, applying that definition and establishing a qualitative assessment of whether a design has a lot or a little is—well—not so easy.

For example, in an absolute sense, the number 100 is neither large nor small. For a mechanical pencil sharpener, 100 interconnected parts would be a lot. Absurdly so. For a jet aircraft, it would be shockingly few. Similarly, if you have a dozen pennies, that's not a lot. But a dozen kids? Wow! Clearly, context matters.

One way to deal with the ambiguity of "a lot" is to introduce the concept of efficiency. A perfectly efficient design has no gratuitous elements. Instead, it has just the right number of interconnected parts, each of which carries its own weight and contributes positively to the overall operation of the system. It operates with minimal friction, effort, and waste. Whether composed of 10 parts or 10 million, an efficient design does not have "a lot" of interconnected pieces. It has just enough. This is not exactly the same thing as minimalism, which carries connotations of extreme sparseness, but it does share a common preference for the "minimal effective dose" rather than overdoing it.

Consider a well-designed vehicle dashboard that conveys only the most important information to the driver at any given time. Vehicle speed is always displayed because that information is always needed, while the temperature light only illuminates when the engine is particularly cold or in danger of overheating. Most of the time, the fuel gauge provides a general indication of how much gas remains, but if the level drops below a certain point, a new indicator lights up to alert the driver of the need to get gas. This fosters an efficient use of the driver's mental resources, only asking us to pay attention to things that actually require attention.

Using the words *simple, complex, efficient,* and *a lot* in this way is a philosophical and literary position, not a scientific one. There are other ways to use these words, and other reasons, so we should be clear about what we're doing here: adopting a relatively simple definition of complexity and connecting it with the concept of efficiency.

Now we turn to the other word in the diagram, the label on the horizontal axis. The word is *Goodness* and we have used that word several times already, but what does it really mean?

For a technical system, it might mean operational functionality, effectiveness, or fitness-for-use. For a book, presentation, or other attempt at communication, goodness probably includes elements of clarity and accuracy, as well as some measure of interestingness. Making a meal? Goodness is a blend of delicious and nutritious, although convenience and cost could certainly come into play as well. For business, goodness is tightly linked to profit, while socially minded companies who pursue a so-called triple bottom line also consider the impact on people and the planet.

It is worth noting that goodness does not mean perfection. It just means goodness, and a thing with flaws might still be quite good. For that matter, in *The Reflective Practitioner*, Donald Schön points out that "descriptions that are not very good may be good enough," while the cost and delay associated with improvements beyond a certain "good enough" point may be not only unnecessary but also unwise.

Using a general term like *goodness* allows us to apply the concept across a wide range of specific situations. Readers are invited to create their own definition of the word, based on whatever particular measures of merit are relevant for their situation. It is worth noting that *goodness* is a relative term whose definition varies over time. As Henry Petroski explains in *Small Things Considered*, "We evaluate designs not against absolutes but against one another." Thus our evaluation of a thing's goodness will not remain static so long as the alternatives continue to change.

With the outline established and the terms defined, we can now begin moving around. But where to start? Well, sometimes we are fortunate enough to begin a design effort with a blank sheet of paper (metaphorically or literally). This is not always the case, but when it happens, we find ourselves at the bottom left corner, in the Region of the Simplistic, as shown in Figure 2. The single square in that corner is a metaphorical representation of our first design element. We'll add to it as we proceed.

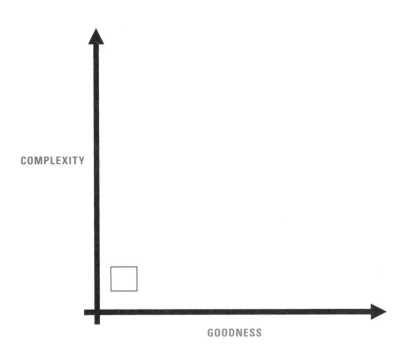

FIGURE 2: THE REGION OF THE SIMPLISTIC

In this region complexity is low, and so is goodness. We haven't created, added, or done very much yet, so our design is neither very complex nor very good. That's all right. Everyone has to start somewhere, and this is where we begin to lay a foundation for all the progress and work that will follow. But it is boring here, so we don't stay long. In fact, it's almost impossible to stay here because of our natural tendency to make additive enhancements.

We leave this area by adding pieces, parts, and functions to the design. Books gain words, sentences, paragraphs, and chapters. Pizzas accumulate sauces, seasonings, and toppings. Software incorporates lines of code, modules, and dynamic-link libraries. Processes expand to include new steps, procedures, reviews, and decision points.

We add these things for a fundamental, benign reason—they make our designs better. Pizza without toppings is just flat bread, but sauce and cheese make it delicious. A book without ink is just blank paper, but words and pictures add meaning. We want to improve our bread and paper, to increase their goodness, so we put things on them.

But these additions do not only increase goodness. They also increase complexity. More words in our book, more modules in our software, more steps in our process, more ingredients on our pizza all contribute to a more complex product.

We can depict this change with a line that moves up and to the right, in the direction of increased complexity and goodness, as shown in Figure 3. I call this the Complexity Slope.

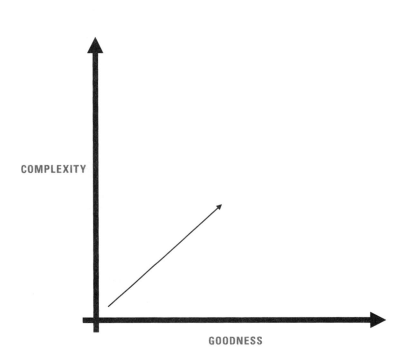

FIGURE 3: THE COMPLEXITY SLOPE

Positive progress along the Complexity Slope can be described as *learning* and *creating*. In a word, the slope is about genesis, the production and addition of new parts. Sometimes the slope is steep, when complexity rises at a faster rate than goodness because the benefit of each addition is small. Other times the slope is gentler, as small increases in complexity produce significant boosts in value. But regardless of the angle, things are becoming simultaneously better and more complex.

As an example of how this might happen, consider the automobile. Once upon a time, cars didn't have seat belts. They didn't have airbags. They didn't have any number of safety features that are now standard on even the least expensive models. These added safety components make today's cars more complex than the cars our parents and grandparents drove, but there is little disagreement that the increased degree of safety also means they are better.

Most seat belts are themselves quite simple. True, some restraints are more elaborate than others, with pads and sensors and such, but even the fanciest version is essentially a strap, a ratchet, and a latch. In the big scheme of things, a seat belt represents a minor increase in complexity for the vehicle.

As for goodness, the value this safety device conveys to the driving experience far exceeds the complexity it adds. The National Highway Traffic Safety Administration says if you're in a crash, wearing your seat belt cuts the risk of injuries by 50 percent. That's a good thing.

This shows that increases in complexity can make the design better. No big surprise there. However, the relationship between complexity and goodness is not constant. They are not *always* directly proportional, and making something more complex only improves it to a point, which brings us to the center of our map.

In the center of the map we find a critical mass of complexity. This area is not exactly the proverbial sweet spot where goodness is optimized—that comes later. Rather, this is an inflection point, a phase shift where the rules of progress begin to change. I call it the Region of the Complex, and it is shown in Figure 4.

At this point, the number of elements involved has substantially increased beyond the original simplistic design, and we have achieved a meaningful degree of functionality and maturity (aka goodness). We have now amassed many design elements, shown here as a collection of squares. We might even say we have . . . a lot of them.

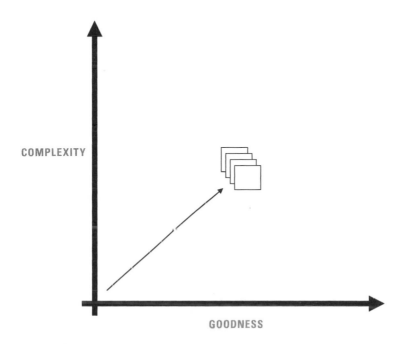

FIGURE 4: THE REGION OF THE COMPLEX

There is nothing wrong with spending a significant amount of time in this area. The designs we find here are demonstrably better than their predecessors and arriving here indicates an admirable degree of progress. We are no longer dealing with early prototypes and first drafts, but neither are we talking about finished, polished end products, either. Instead, designs in this area are intermediate versions that may be sufficient to put into the hands of our beta testers and early adopters, but are unsuited to be fully released into the wild.

In the early 1900s, the Wright brothers landed their first plane squarely in this area. The Wright Flyer was a rather complex machine and required a fair amount of effort and maintenance to keep it aloft. It was no longer a scale model or a paper airplane, and it was able to actually carry a person into the wild blue yonder. But it was not quite ready for sale to the general public just yet.

The Wright Flyer was simple by today's standards in that it lacked such niceties as a cockpit, a radio, and retractable landing gear, but its creation was primarily the product of genesis and learning. Orville and Wilbur busily produced new information and new functions and added to earlier designs. Arriving at this point required some simplification and integration, to be sure, but much more of that was still waiting in the future.

For example, the double wings and exposed struts they used represent a level of complexity not found in later improvements. Interestingly, some of Orville and Wilbur's antecedents used as many as five sets of wings, but modern aeronauts agree a simpler, single wing design is preferable.

As the concept of an airplane matured and engineers came to a more complete understanding of what flight truly requires, several aspects of its design were simplified and removed. Other aspects got more complex. We'll address this apparent paradox shortly.

While it is perfectly fine to linger here in the center of the chart, learning and exploring and modifying our design, at some point we'll need to make the types of changes that transfer us to a new region. Maybe a competitor's arrival will inspire us to revamp our design, or maybe we just get bored and want to explore something new. Regardless of our motivation for packing up and moving out, we now have a choice to make. There are two paths out of this central region and neither follows the earlier trajectory of increases to both complexity and goodness. We have to make some changes in our approach, and as the Grail Knight said to Indiana Jones in *The Last Crusade*, it is important to choose wisely.

Note that it is not possible to continue moving up and to the right indefinitely, increasing both complexity *and* goodness in order to produce something that is simultaneously supergood and supercomplex. That means the upper right corner of the diagram is Strictly Off-Limits. Like certain parts of Maine, you truly cannot get there from here . . . or from anywhere.

Instead, the two departure paths run perpendicular to the path that brought us to the middle of the chart. One heads up and to the left; the other heads down and to the right. Figure 5 shows what that looks like.

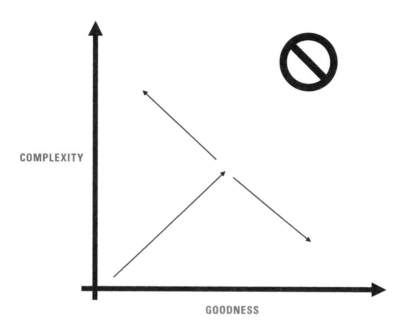

FIGURE 5: TWO PATHS AND THE OFF-LIMITS AREA

Let's first examine the path we want to avoid—the upper path—before turning our attention to the desirable lower path. This upper path is the one we'll follow if we continue exhibiting the same design behaviors that brought us to the middle of the chart—adding, creating, and expanding.

As we pile on more and more layers of complexity beyond

this central inflection point, our design gets worse. Goodness decreases. Even though our increases in complexity are done in the name of improvement, the result is a deterioration of performance, reliability, and quality.

You have probably heard of the Law of Diminishing Returns, where each new addition or change conveys less benefit than the previous one. Yes, it's a bummer when the rate of improvement slows down, but what's going on at this point is even worse than that. Instead of diminishing returns, this upper path depicts *negative* returns. Along this slope, new additions do not merely provide less value than the ones before. They actually weigh down the design and make it worse. The rate of improvement hasn't slowed—it has reversed direction entirely.

The Germans have a wonderfully long word for this type of "improvement" that makes things worse: *verschlimmbesserung*. That's precisely what's happening here. The more we tinker and tweak, the more we add and expand, the worse things get. Adding to our frustration and confusion is the disconcerting knowledge that our additive behaviors used to be productive.

What changed? Our location on the map changed. We found a bend in the trail and instead of heading east are now going north. It may feel like we're on the same trail as ever, and we may be wearing the same well-worn hiking boots, but now the path is overgrown, the sun is setting, and it's starting to snow. Plus, we're being chased by a mountain lion. And a grizzly bear. And we're out of coffee.

Our situation has changed enough that we can give this route its own name: the Complication Slope. You can see it in Figure 6. Movement along this slope is the result of overlearning and smugness, where we fall in love with complexity for its own sake rather than using complexity as a means to an end. We mistakenly see increases to complexity as signs of progress, and continue certain design behaviors even after they have ceased to be productive.

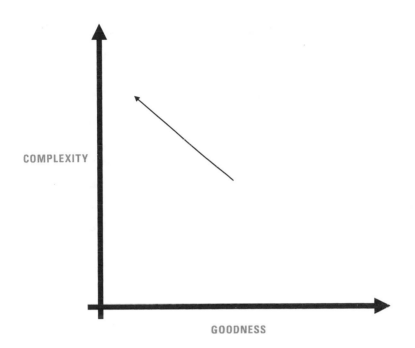

FIGURE 6: THE COMPLICATION SLOPE

There are several reasons we might move a design along this unfortunate path. Sometimes it's just a matter of intellectual inertia, where we get so used to adding to the design that we keep doing it even after such additions cease to be useful. Other times we do it because we're quantifying and rewarding the wrong thing, treating complexity as a measure of merit or a desirable design attribute, as if customers prefer buying things they can't understand.

In a 2006 *BusinessWeek* article titled "How Do You Turn On the #@!&% Air?" David Welch bemoaned the complicated dashboards of luxury cars that require navigating through multiple menus to perform simple tasks such as adjusting the temperature. He poignantly asks, "Whatever happened to the button with the snowflake on it?" Digital control panels may look fancy, feel posh, and help justify large price tags, but they don't necessarily improve the driving experience. In fact, they just might make things worse.

A commentator for the Innovation-TRIZ newsletter shed further light on the dashboard problem, writing, "Isn't it fun to add things? Don't we engineers love to design something, to add to something, to control something? In hundreds of . . . workshops using simple introductory problems, when a group of engineers is asked to create a solution to a problem, 99% of the time they invariably ADD something to the system to fix the problem."

It's unlikely all those additions represent improvements. It's even less likely that 99 percent of design challenges require something to be added. And yet, addition is the default mode for engineers (like myself). A more thoughtful approach reveals that design problems are sometimes rooted in what is unnecessarily present rather than what is missing. In such cases, the solution involves a removal, a reduction, a subtraction, a decrease . . . and if it weren't so busy serving as a bad example, I think the previous sentence would be better if it was less redundant.

The underlying dynamic is rooted in the fact that while complexity has a value, it also has a cost. As Darrell Mann explained in an article for the *TRIZ Journal*, eventually "the problems that come with the increased complexity outweigh the benefits."

Poet Cliff Crego illuminates this situation brilliantly when he writes, "complication leads to contradiction." He verbally paints images of gears grinding against each other, and he challenges his readers to face up to the costs involved with avoidable complexity:

The way of thinking which is the most inappropriate of all is the one which does not see, or worse, simply tolerates, unnecessary difficulty.
From there, it is only a small step to the degeneration of cults of collusion which not only condone, but actually cultivate complicatedness.

Membership in these ridiculous cults is easy to come by, and sometimes even the wisest of us join the movement without deliberately deciding to sign up. It starts when we view complexity as unavoidable and inevitable, creating mental inertia that might even lead us to see complexity as desirable. What follows is a pattern of behavior in which we make things more complicated in the hope that such complexities will make things better, ignoring the mountain of evidence to the contrary.

Examples of such behavior abound, but let's consider one that has a happy ending. In 1893, Whitcomb Judson was granted a patent for a device he called "a clasp locker." It was emphatically ignored by the public when he demonstrated it at the 1893 World's Fair in Chicago, but eventually his idea caught on and today we know clasp lockers as zippers. According to the account in Henry Petroski's *Evolution of Useful Things*, Judson's friend Colonel Lewis Walker observed that "Judson's way of meeting a difficulty was to add invention after invention to his already large supply. . . . Judson's activities were expensive. They tended to create more problems than they solved."

The negative impact of Judson's tendency toward complexity, in terms of both dollars and headaches, was obvious to Walker. For that matter, Judson himself was surely aware of it, too, even without Walker's critique, but he persisted nevertheless in his additive approach to problem solving for quite some time. He eventually broke out of that pattern, removed some of the complexifying snags from his design, and changed the face of fasteners forever.

Judson's story is hardly unique. When all the complexity we add to our design fails to bring the hoped-for improvements and instead introduces a host of new problems, many of us have a tendency to resolutely attack those problems by adding even more layers of complexity. We may even go so far as to criticize our benighted customers for lacking sufficient sophistication to appreciate the genius of our design. If drivers would only read the three-hundred-page manual and then attend our convenient four-hour introductory class, they would easily learn the seventeen-step process for turning on the air-conditioning. And if they sign up for the two-week Intermediate Training, they can also learn how to turn the air conditioner off. Don't blame the designer if our customers are too lazy to put in the effort, right?

This is not a particularly effective design strategy, nor a wise one. The best I can say about it is that it does not require much imagination or skill. And I should admit I'm just as prone to going down that path as anyone, as the following story shows.

My fellow graduate students and I were excited when the professor handed out Lego Mindstorms kits. The assignment: build an autonomous robot vehicle capable of navigating a short maze. We'd been looking forward to this project all semester.

All we had to do was write a short list of functions required to maneuver from Start to Finish, then assemble the small collection of sensors, wheels, and blocks into a coherent vehicle capable of performing those functions.

That is precisely what most of my classmates did. Unfortunately, my partners and I took a different approach. As we made our list, we didn't pass up a single opportunity to make things complicated. If a function could be performed with one step or one piece, we used eight.

Our resulting monstrosity was more than twice as large as most other contenders, and it worked even worse than you might expect. On Demonstration Day, we failed to complete the maze on our first run. And our second run. And third. After watching our more successful classmates, we did a last-minute redesign, stripped out a lot of unnecessary junk, and finally made it to the finish line. Not exactly my finest moment.

The team that got through the maze fastest (on their first attempt, naturally) had the simplest and smallest device of all. It used just two wheels and a single sensor. Their software program was a few lines long.

That team may have received a better grade, but I'd like to think my partners and I learned more.

Stay on the complication slope long enough, continually adding unnecessary pieces and *vershlimbesserung*-ing along, and we will arrive at the Region of the Complicated, in the upper left corner of the map, shown in Figure 7. *Hic sunt dracones.* Instead of more squares, we now have meaningless squiggles and pieces that do not (and cannot) fit with other pieces. It's a mess up here.

When I find myself floundering around here, I am reminded of the immortal words of *Arrested Development* character Gob Bluth: "I've made a terrible mistake." Somewhere along the way I managed to confuse complexity with value and ended up making things worse instead of better.

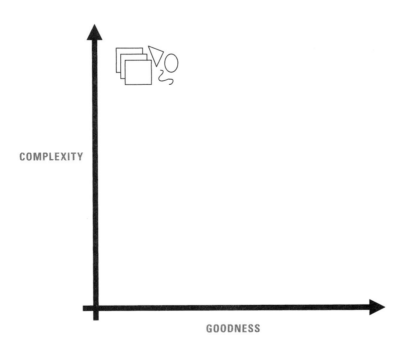

FIGURE 7: THE REGION OF THE COMPLICATED

This area is markedly different than the Region of the Complex, found in the center of the map. There's nothing wrong with hanging out there. There is everything wrong with spending time in the Complicated area.

Complex and *complicated* may sound similar, but they are in fact very different beasts. Complexity is often both essential and unavoidable. As simplicity guru John Maeda famously wrote, "Some things can never be made simple." Certain topics, issues, activities, and missions are inherently *complex*—and that is okay. Complicatedness, on the other hand is less admirable. Complications arise from *unnecessary* complexity, from the addition of non-value-added parts, of gears that turn without reason or grind against other gears. Complexity may be essential, but complicatedness is not.

What does unproductive complexity look like? Well, a few years ago, several media outlets excitedly reported a breaking news story that the U.S. military had unexpectedly published an unnecessarily complicated set of specifications for a piece of gear. Students of military history know this was not an entirely unprecedented situation, but the example in question was interesting enough to merit a headline or two.

Was the story about a high-tech stealth fighter? A particularly massive aircraft carrier? An amphibious rocket-submarine? Nope, nope, and heck no. The reports were about a baked good.

Turns out the Army had created a twenty-six-page manual on how to bake a military brownie for inclusion in an MRE (meal, ready-to-eat). To be fair, the manual also addressed oatmeal cookies, and several pages dealt with the packaging requirements, but everyone seemed to agree twenty-six pages was a bit much.

Of course, we aren't talking about ordinary brownies. These

combat-trained treats have a shelf life of three years and are regularly subjected to hostile environments that would be fatal to their more delicate civilian counterparts. But as someone who has been on the receiving end of this confection's sweet attentions, I can confirm it has almost nothing in common with the yummy, chocolaty dessert we all know and love. On the other hand, I shouldn't complain too much about them. There have been times when eating one was the best part of my day, and I know I'm not alone in that experience.

While MRE brownies often serve as bright spots in otherwise dark places, it is not unreasonable to ask whether the recipe's complexity was necessary, productive, and good. Most likely it was not. Surely the instructions could have been done more simply, and the simplicity might have improved the outcome, if not the taste.

As an alternative data point, my grandmother put her brownie recipe on a three-by-five card. Then again, unlike the military version her brownies never lasted much past dinnertime.

How do we end up with a twenty-six-page brownie recipe? More generally, how does complexity become complication? What triggers the tipping point, where positive additions become negative, where design strategies that were previously sound become downright destructive? It starts when we overvalue complexity, equating it with sophistication and implicitly assuming that if a little bit of complexity is good, then a lot is even better. But it also happens because changes in quantity can indeed change quality. Our friends in the field of software development can shed some light on this phenomenon, which they sometimes call the State Explosion Problem.

The State Explosion Problem goes like this: as the number of variables in a computer program increases, the number of possible states the program can assume increases exponentially. In noncomputer terms, the more pieces we have, the more possible interactions there are between those pieces. Lots of pieces means a mega-lot of interactions.

Why is that a problem? For starters, we may want to test the program's performance before sending it out into the world, to see if it does what it's supposed to do. If the number of possible states has exploded exponentially, there simply isn't enough time left in a dozen lifetimes to test each scenario. If we can't test the design, we can't know how it will perform in real life. That can lead to all sorts of nasty surprises.

Excessive complexity not only makes testing difficult, it also makes designs more fragile by increasing the number of possible failure modes. That is, each piece we add introduces new opportunities for breakage. Accumulate enough potential break points and one of them is bound to give way. That's not good.

Along with increased fragility, there seems to be a correlation between the upper left quadrant, with its confusing and upsetting complexity, and a certain degree of mental distress. I wonder: Does complexity drive us crazy? Or does causality flow the other way? Perhaps insanity drives us toward complexity.

In much the same way, the profound and comforting simplicity found in the lower right corner goes along with that most desirable element of mental health, serenity. Is inner peace a root or a response? Do we enter a state of Zen when we focus on creating elegant simplicity in our designs? Or do we produce simplicity because we first possess internal serenity?

I suspect our mental state is both a cause and an effect. The relationship between (in)sanity and (un)clarity works in both directions. Thus, being unfocused and frantically confused degrades our work and fosters unnecessary complexity in our designs, as we flail around and add components in the blind hope that some of them will make the design better. This reflects back on our inner lives and causes more confusion, an unfortunate vicious circle.

Likewise, when we are internally centered and focused, we can more easily avoid getting our design wrapped up in entangling complexities because our vision is clearer. We see and embrace alternatives that get overlooked or dismissed when we are in a more harried, hurried state. This flows back into our core, as the presence of simplicity in the externals of one's life contributes to a sense of serene well-being in one's inner, mental life.

That's why I ran out to the store last week and bought one of those mini Zen rock gardens, plus a tabletop waterfall fountain (with soothing LED lights built right in), a dozen candles in various sizes/colors/scents, a white-noise generator, and an aromatherapy diffuser. The only problem is I now need a larger desk, because my current desk does not have enough space for all that stuff plus my laptop. Hold on, I'll be right back.

The most famous documentarian of this relationship between complexity and insanity was Rube Goldberg. His famously convoluted drawings shone a humorous light on humanity's tendency to overvalue complexity far beyond reason. He described his creations as "symbols of man's capacity for exerting maximum effort to accomplish minimal results."

Goldberg's devices clearly reside in the upper left corner of the map and demonstrate the irrationality of that area. Unlike most residents of that region, these ended up there on purpose. They deliberately accomplish in five steps a task that should only take one, and they do it to make a point. It's funny when they do it on purpose. It's less funny when we do it inadvertently.

Despite the long-standing popularity and widespread awareness of Rube Goldberg's pointed commentary on complexity, some people still manage to believe complexity is always virtuous. The mental gyrations necessary to sustain this position look positively exhausting, but I've got to admire their commitment.

One such attempt to defend the importance of high levels of complexity occurred in a 2013 article published in a learned academic journal whose name I would rather not share. The authors of this particular piece asserted without any hint of irony that program managers and engineers never "add unnecessary complexity to their systems and processes without reason."

Unlike Goldberg's work, I suspect the humor and the truth in that journal article were both unintentional, but I found myself laughing and agreeing with the sentiment all at the same time. Yes, people always have reasons for the unnecessary complexity they add to their work. Rationalizations abound. However, unnecessary complexity is, by definition, *unnecessary* and the reasons provided are entirely unreasonable. The result is confusion, complicatedness, and expansive bureaucracy that aims to cope with unnecessary complexity by adding more complexity, on the grounds that the only way to address an unreasonably high level of complexity is by adding to it. Surely not.

It perhaps bears repeating that a certain amount of complexity is valuable and necessary, but at some point complexity devolves into unreasonable complication, which makes things worse.

Interestingly, the dynamic of complexity becoming complication occurs even when the thing we're designing isn't a thing.

Throughout his book *Ambient Findability*, information architect Peter Morville discusses the problem of information overload. He argues that in addition to being painful and distracting, too much information actually leads to worse decisions.

We could depict the relationship Morville describes between information quantity and decision quality as a bell curve, with decision quality rising, then falling, as we gather more and more information. At first, gathering additional facts leads to better decisions. But at a certain point, learning more actually reduces decision quality. This pattern should sound familiar by now: facts and details increase complexity and help us make better decisions, until we go too far and the overwhelming complexity makes things worse.

Morville's observation suggests that facts and details are good and important, but they can be difficult to manage in large quantities. One way to manage them is by transforming facts into metaphor and story. This allows us to integrate them into a simpler block of knowledge that is easier to understand, remember, apply, and share.

Metaphors give us a handle with which to grasp the essential facts. Metaphors also make it easier to pass along the facts to a fellow traveler. For example, at the risk of being too self-referential, we might talk about design as a journey and designers as travelers. This creates a framework for additional discussion, which may include things like "starting out," or "a bend in the road," or even "a map of complexity-related pitfalls." Without some sort of metaphorical structure, our conversation would quickly be overwhelmed by the weight of precise, literal descriptors that are much harder to grasp and apply.

Fortunately, using metaphors comes naturally to us. Whether we are trying to be metaphorical or not, our minds default toward metaphor automatically, sometimes without us even realizing it. In fact, George Lakoff and Mark Johnson's book *Metaphors We Live By* argues that metaphor is a "fundamental mechanism of mind," and that we can scarcely think of anything in nonmetaphorical terms.

In a similar vein, Marshall McLuhan wrote, "When you give people too much information, they instantly resort to pattern recognition to structure the experience." This is not exactly a choice. It's just the way humans are wired. It's how we make sense of the world around us—by replacing complex literalism with simpler, metaphorical descriptions. We describe one thing in terms of something else, a metaphor or a pattern, to make it easier to handle.

Not all metaphors are created equal, and some are more useful or enlightening than others. Since we're going to use metaphors anyway, we should make an effort to construct our metaphors thoughtfully and ensure they illuminate the key facets in question and don't obscure any critical elements. In doing so, we will avoid the overwhelming weight of information Morville warned us about.

Whether the topic is information or hardware, excessive complexity is often rooted in a lack of priorities. When we can't tell which pieces of information or which aspects of our design matter, we try to include everything for fear of missing something important. The result—we miss almost everything.

The fear of missing out is not unfounded. Yes, in the name of simplicity, we might miss something. So choose priorities carefully. Thoughtfully. Even perhaps prayerfully, or at the very least meditatively and reflectively.

But choose.

This does not mean we should post a cloyingly vapid mission statement on the wall, helpfully informing everyone that profit/quality/safety/excellence is Job No. 1. That superficial approach to prioritization fosters cynicism. What I'm talking about is more practical than that, as the following story shows.

Software developer Joel Spolsky wrote a fantastic essay on the foolishness of unnecessary complexity, titled "Choices= Headaches."* He described discovering fifteen ways to shut down

---

* *www.joelonsoftware.com/items/2006/11/21.html.*

a laptop running Microsoft Vista, including sleep, hibernate, Switch User, four different function key combinations, closing the lid, and, of course, the on-off button.

There was probably a sound reason for adding each method, but there is no sense in having all fifteen.

Some might object that extra features are harmless and don't get in the way, but that is untrue. It's also answering the wrong question. See, it's not enough for each function to be merely harmless. We should instead ensure every component makes an actual contribution and improves the design. Every piece should be useful, and as Spolsky pointed out, too many "harmless" additions cause headaches.

Designers therefore should ruthlessly scrub out unnecessary redundancy and mediocrity, no matter how well-intentioned. That may sound obvious, but Microsoft's Vista Power Down Committee apparently skipped that step.

Fortunately for us all, Microsoft learned its lesson and took a different approach to Windows 7. Instead of blithely adding a dozen different ways to achieve the same effect, they focused their design effort, rewarded coders for streamlining the software, and produced a simpler, smaller, less buggy product.

Removing the headache-inducing components from our design means we are moving along the Simplification Slope. As Figure 8 shows, this means our design now moves down and to the right rather than up and to the left. In order to move along this slope we have to learn to use some new tools . . . and set aside some old ones, at least for now. In place of learning and genesis, which served us well along the first part of the trip, we must now adopt a tool set that includes things like unlearning, synthesis, and reduction.

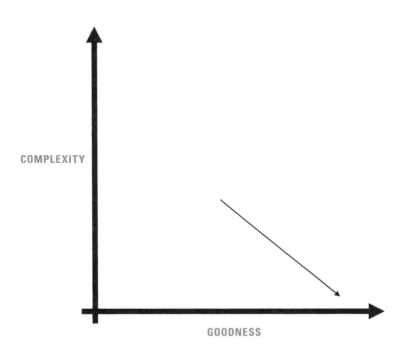

FIGURE 8: THE SIMPLIFICATION SLOPE

Along this slope, the most productive tasks do not involve creating new elements. Instead, we must integrate the existing elements and creatively discard the unnecessary parts. The idea is to prune and pare down the design, reducing it to the essential components, each of which is able to freely operate with minimal friction and maximum contribution.

This 90-degree shift in behavior is harder than it sounds, in large part because it demands we overcome a great deal of design momentum. When we've spent a lot of time adding things to the design, the transition to a subtractive approach can feel uncomfortable, awkward, or even wrong. It also requires a certain amount of humility as we remove things we'd previously created. Despite the difficulty, it is worth the effort.

Speaking of difficulty, having a taste for simplicity is not the same as having a talent for it, and our preferences inevitably precede proficiency. That is, we want simplicity long before we are able to achieve it, and our early efforts to create elegant simplicity seldom live up to our expectations. This leads many beginners to throw in the towel prematurely.

Michael Caine's con man character in the 1988 film *Dirty Rotten Scoundrels* blamed the disparity between skill and ideals for his wayward lifestyle. He explained it to Steve Martin's character in these words: "Freddy, as a younger man, I was a sculptor, a painter, and a musician. There was just one problem: I wasn't very good. As a matter of fact, I was dreadful. I finally came to the frustrating conclusion that I had taste and style, but not talent." Abandoning the effort to create, he instead used his exquisite taste to liberate large quantities of money and treasure from the wealthy. Hilarity ensued.

Public radio host Ira Glass tells much the same story, minus the criminal element: "Your taste is why your work disappoints you. A lot of people never get past this phase, they quit. . . ." Glass goes on to explain the key to overcoming this talent deficit: "It is only by going through a volume of work that you will close that gap, and your work will be as good as your ambitions." In other words, practice and persistence turn preference into proficiency.

Yes, the transition toward simplifying a complicated design can be difficult. One of the best ways to start down this new path is to take a pause from our previous path. Stop adding. Stop creating. Stop everything.

Just stop.

Bear in mind, this suspension of activity is a temporary halt, not a permanent one. The idea is to pause, not quit. There is still plenty of work to be done. And the remaining work is precisely why we need to take a breather. As leadership guru Kevin Cashman wrote in *The Pause Principle*, a pause is "exactly what is needed to sort through complexity and then drive performance to the next level."

In this pause, go do something else. Walk outside. Visit a friend. Sing a song. Paint a picture. Doodle. Exercise. Meditate. Feed the soul. Whether it's for an hour or a year, there is much wisdom in taking a sabbatical at this particular point in the design process, as it breaks up our momentum before it carries us somewhere we don't want to go. There is also scientific support for this practice, particularly if the thing you decide to do is take a walk.

Stanford researchers Marily Oppezzo and Daniel L. Schwartz published a paper in the *Journal of Experimental Psychology* in 2014 exploring the effects of walking on creativity. They performed several experiments that "demonstrate that walking boosts creative ideation," and they observed that "[w]alking opens up the free flow of ideas, and it is a simple and robust solution to the goals of increasing creativity and increasing physical activity."

Walking is great, but when I need a pause, I prefer to juggle. Juggling can be invigoratingly athletic or quietly meditative, depending on my mood and whether I need to get the blood pumping or to center my breathing. Regardless of the pace, juggling has a physical effect, as I move my arms and focus my eyes in a new way. It also has a mental effect and allows the problem-solving, design-oriented, Type-A part of my mind to slip into the background while the younger, more playful part takes center stage. This engages the creative subconscious mind, which comes up with ideas, approaches, and solutions my more rational conscious mind failed to consider.

I keep a set of juggling balls on a bookshelf in my office for this very reason, and I've found that even five minutes can provide an effective reset, breaking the additive momentum and setting the stage for a new, simplifying approach. Sure, it attracts some funny looks from my colleagues sometimes, but that's a small price to pay. For that matter, I kind of like getting those funny looks.

Speaking of colleagues, I've also found that learning to juggle is easier than it looks. I can usually teach someone how to juggle in less than half an hour. Mastering the skill takes a bit longer, but the basics are easy enough to grasp over a short lunch break, with plenty of time for a sandwich afterward.

Some people prefer a more sedentary type of pause. Winston Churchill's habit of taking daily naps is well known and widely documented. In his own writings, Churchill explained how a nap helped "renew all the vital forces" and enabled him to maintain a prodigious output as a leader and statesman. Nappers thus are in good company, and although I've never developed a taste for daytime sleeping myself, I unreservedly recommend naps as a wise thing to do when you need a pause.

Sleeping in the middle of the afternoon was not the only type of pause Churchill practiced. At the age of forty, he began to paint for the first time, going on to win several awards for his artwork. In a brief essay titled "Painting as a Pastime," Churchill explained the importance of cultivating a hobby like painting as a way to renew and refresh one's mental faculties, to experience a pause that genuinely refreshes. He wrote, "Painting is complete as a distraction. I know of nothing which, without exhausting the body, more entirely absorbs the mind."

The specific medium does not matter all that much, although Churchill strongly preferred oils to watercolors (I prefer sketching in pencil, in case you were wondering). The all-absorbing aspect of the activity is what matters most. With a brush in his hand and a landscape before him, Churchill was briefly free from the unrelenting demands of statesmanship. These sessions rested his "mental muscles," to use his phrase, and he returned to his duties with new vigor.

The heart of Churchill's "Pastime" essay is captured in the following advice: "To be really happy and really safe, one ought to have at least two or three hobbies, and they must all be real." Whether we decide to adopt painting, juggling, napping, or something else, there is great value in having a handful of hobbies we can turn to when we need a pause.

A pause is an opportunity to reorient. To become reacquainted with our objectives, our priorities, our desires, and our obstacles. To become more mindful and less frantic, to interrupt the creative momentum before it carries us away. And to prepare for the return.

When we resume the work, we bring new clarity with us. We come with fresh eyes and new questions, as we reapproach a design that now seems strangely alien and unfamiliar. It has not changed. We have.

"What does this piece do? Why is it here? Do I need it?"

Need a pause now? Go ahead and take one. Go for a walk, do a little juggling, or grab a nap. You may even want to grab a pencil and fill in the margins and blank spaces of this book with doodles. I don't mind—it's your book, after all. So let your mind wander as you draw. When you're ready to return from your pause, turn the page.

CHAPTER 3

The Journey Continues

Welcome back. I hope your pause was a good one, and I hope it becomes a regular part of your creative practice. Speaking of regular contributions to creativity, we begin this chapter by introducing a gentleman named Genrich Altshuller.

Altshuller was an inventor, engineer, and writer living in the Soviet Union. In 1950, Joseph Stalin threw him into the gulag during a political purge, along with a bunch of other "dangerous" intellectuals. Fortunately, he survived and was released four years later. His exposure to so many brilliant minds during this incarceration helped him create the *Theory of Inventive Problem Solving* (aka TRIZ, the theory's Russian acronym), which is now used by engineers, inventors, and designers around the world.

TRIZ offers practitioners a remarkably powerful toolbox, full of useful practices and principles of invention. The flowchart in

Figure 9 illustrates one particular TRIZ practice called "trimming." In a nutshell, trimming involves removing a part from the design, then using the remaining parts to perform the necessary functions. Selecting the part to trim out can be arbitrary or deliberate. We may start by removing a part that is obviously extraneous. Then, for a fun challenge, try removing a part that looks essential.

Trimming can be used in a wide variety of situations, from writing code to crafting a presentation or an organizational structure. In Simplicity Cycle terms, it helps us move down and to the right, along the Simplification Slope. The objective of trimming is to remove the contradictions, smooth out the friction, and erase unnecessary redundancy. How do we know when we're done? As Antoine de Saint-Exupéry explained in *L'Avion*:

Perfection is achieved not when there is nothing more to add, but rather when there is nothing more to take away.

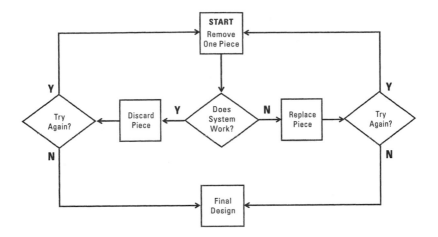

FIGURE 9: TRIMMING FLOWCHART

As an example of what trimming looks like in practice, let's look at one of the last research and development projects I led. The Dismount Detection Radar system, affectionately known by the catchy name DDR, is a cigar-shaped pod containing a radar antenna and assorted supporting components. It is designed to be attached to the underside of a military aircraft and flown over hostile territory to hunt for bad guys traveling on foot (that is, "dismounts"). It is a pretty cool-looking piece of hardware, even though my repeated requests to paint shark teeth or a handlebar mustache on the front kept getting denied. Nobody appreciates good nose art anymore. But I digress.

The electronic components inside the DDR pod are sensitive to extended exposure to heat and thus must be cooled to prevent damage, particularly when operating in a hot environment. During our flight tests, as the aircraft sat on the desert tarmac prior to takeoff, we connected the pod to a large tube that pumped cold air into the system and kept the inside nice and frosty. The cooling unit was larger than the pod itself and there was no way we could haul it up to thirty thousand feet. Fortunately, we didn't have to.

Despite what the myth of Icarus tells us, flying closer to the sun does not expose us to hotter temperatures. In fact, as we head off into the wild blue yonder the ambient air temperature drops precipitously, so the wax holding Icarus's feathers in place is more likely to freeze than melt. This means that while the ground crew is enjoying a blistering summer day on the runway, the fliers are dealing with temperatures approaching minus 30 degrees Fahrenheit. That's great news for DDR's heat-sensitive hardware. It's also great news for the system's overall size, weight, and complexity.

Instead of permanently installing a big gas-powered air conditioner on my streamlined little radar pod, which would have made the system heavier, more expensive, and more complicated, all we had to do was get to altitude quickly and allow the air intakes to carry deliciously cold air throughout the pod. Yes, the cooling trailer was essential prior to takeoff, but once we got clearance from the tower we trimmed it away and used a simple scoop and duct system to provide the same function.

Trimming comes in many flavors, and DDR illustrates a specific type of time-shifted trimming where the mechanism in question is included during some portions of the system's operation, but removed during other portions. In other applications, the trimmed component might be permanently replaced with a simpler mechanism or even discarded altogether.

Along with clever ideas like trimming, TRIZ practitioners talk about the *Law of Ideality*. This law states that as designs mature, they tend to become more reliable, simpler, and more effective—more ideal.

The Law of Ideality explains that the amount of complexity in a design is a measure of how far away it is from its ideal state. In fact, upon reaching perfect ideality, the mechanism itself no longer exists. Only the function remains. Sort of like doing away with the cooling trailer and using the atmosphere instead.

When the mechanism no longer exists, it has no parts. No

parts equals no complexity. Just pure, simple function. Zero complexity, optimal goodness.

I get all Zen just thinking about it.

Unfortunately, some people believe less complexity automatically equals a worse design. They resist trimming in order to avoid the negative slope shown in Figure 10.

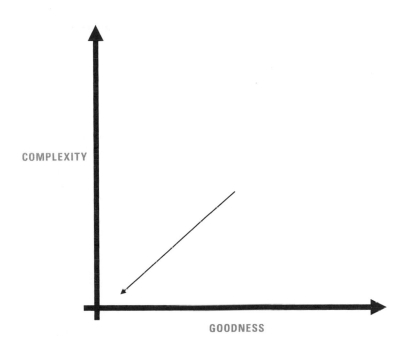

FIGURE 10: THE NEGATIVE GOODNESS SLOPE

Resistance to simplification is based on the belief that every additional feature, part, and function represents an improvement. It also assumes that the accumulated additions made things better from an overall system perspective.

Such assertions are misguided and maybe a little arrogant.

Why "arrogant"? Because they assume everything we ever

added was a good idea. Why "maybe"? Because those additions may indeed have been good ideas at the time. But getting rid of them might be even better.

Even when a new feature makes our design better, some features are "less better" than others. We add something, then another something, then still more somethings. The design improves, but just barely. Our rapid increase in complexity creates a minor increase in goodness. This can be deeply frustrating, as each new addition fails to deliver the degree of advancement we seek. We may take some comfort in the fact that our additions aren't making things worse, but that doesn't help much if we are trying to aim high and change the world.

And then, sometimes, something magical happens. A piece arrives that I call the Special Piece. It's the piece we've been looking for all along without quite knowing what it looked like or even whether it existed at all. It introduces a game-changing improvement, making things far better than any previous addition did and bending our Complexity Slope almost perpendicular to the original trajectory. This final addition single-handedly takes our design to a whole new level of excellence. What a relief!

But our work is not yet done. With the Special Piece in place, we just might discover that many of the earlier pieces—these complexifying additions we sweated and grunted and labored under, with such underwhelming results—are now deadweight. Their whole purpose was to get us to the Special Piece, and now their purpose has been fulfilled, making them prime candidates for removal. And so we begin to—carefully—extricate ourselves from the jumble. Figure 11 shows what that path might look like.

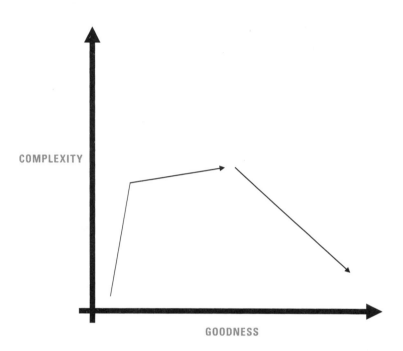

FIGURE 11: THE SPECIAL PIECE

In late medieval architecture, a Special Piece called a flying buttress provided great load-bearing capacity with a lighter, cheaper structure than earlier designs. Outer walls had previously been thick and massive to stand up to lateral strain, but thanks to this brilliant innovation the walls could now accommodate large stained-glass windows, adding to the beauty and function of countless Gothic churches.

More recently, the Special Piece played a prominent role in the development of a funny little boat named *Honey Badger*, which is designed to sail around the world without a crew on board, among other missions. I've never tried to do that myself so I don't know for sure, but I imagine it's even harder than it sounds.

Wired magazine ran an article about the ship and its creators, Richard Jenkins and Dylan Owens, providing a behind-the-scenes look at their design process. Their progress through a series of versions and design challenges makes for a fascinating story, and also shows exactly what a Special Piece can do.

Instead of a fabric sail, *Honey Badger* is propelled by the same type of rigid wingsail found on racing yachts like America's Cup winner *USA 17*. However, the Special Piece is not the wingsail itself. It's the tail that sticks out from the back of the sail. The thing that is so special about this little addition is the way it allowed a radical simplification of the rest of the boat.

While at sea, the boat's autopilot controls a small tab on the back of the tail, automatically orienting the vessel to take best advantage of the wind. Because of the tail, *Honey Badger* does not need the ropes, winches, and crew that would normally be required to adjust and maintain the sail's orientation. The *Wired* writer explains, "By severing all the ropes that run between the boat and the sail on a normal yacht, a lot of the complexity of sailing goes away," and goes on to observe, "Its tail simplified the process of sailing so much that even a robot could handle it."

According to the *Wired* article, "The tail was the breakthrough idea that got Jenkins in the record books, it's what got *Saildrone* [*Honey Badger*'s other name] to Hawaii, and it's what has the potential to disrupt a multi-trillion-dollar slice of the global GDP."

This is classic Special Piece behavior. Adding the single component was a significant improvement in the craft's performance and allowed the deletion of other pieces that had previously seemed essential. The net effect is a drastic reduction in complexity and a huge boost to goodness, quickly moving the design down and to the right, along the Simplification Slope.

Our design efforts don't always follow this path, of course. Sometimes there is no Special Piece and our progress looks nothing like Figure 13. Instead of major improvements accompanying an addition or subtraction, sometimes the big shift happens when we put the existing pieces together in a new and creative way.

The word for embedding one thing inside another or consolidating multiple elements into a single component is *integration*.

As an example, consider the CAPS LOCK button on a keyboard. On its own, it provides a useful function. But many keyboards do not show whether or not they are in CAPS mode, occasionally resulting in the inadveRTENT USE OF CAPS when a clumsy typist accidentally hits the wrong key.

To remedy this, some keyboard designs include a light on the keyboard that illuminates when the CAPS LOCK key is engaged. This increase in complexity aims to improve the design by providing a visual signal of the keyboard's state. Unfortunately, the light is often far removed from the key itself.

A more enlightened designer might embed the indicator directly into the interface, creating a key that lights up when pressed. This integration is both better and simpler, providing useful information with a minimal number of components.

The late Colonel John Boyd loved to present audiences with an integration challenge that went something like this:

Imagine the following situations: cruising across a lake in a boat, riding a bicycle on a sunny day, moving mounds of earth with a bulldozer and taking a vacation in the Rocky Mountains.

Now, throw away everything from the boat except the outboard motor. Next, throw away everything from the bicycle except the handlebars. From the bulldozer, keep only the treads. And from that lovely vacation, keep only the skis.

What does that leave you with? What can you make with a motor, skis, handlebars and treads?

At first glance, Boyd's stack of components appears to be a jumbled pile of unrelated pieces. And yet, with a little imagination they can be assembled into a coherent whole. Boyd's answer? These parts can form a snowmobile.

This little thought experiment is precisely the sort of thing necessary to produce streamlined, simple solutions. Throw away the unnecessary components. Put the remaining pieces together in interesting ways. Make a snowmobile.

The kitchen illustrates another side of integration. Specifically, the untidy side. Breaking eggs is messy work, turning what used to be a smooth, self-contained package into a jumble of shards and a puddle of goo. It does not look like progress.

Don't stop there. Add more pieces—some ham, some peppers, some onion, maybe a little cheese. Then add heat.

The jumbled mess begins to cohere and a delicious breakfast emerges, greater than the sum of its parts. See Figure 12.

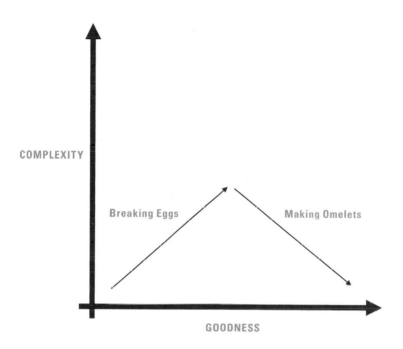

FIGURE 12: EGGS AND OMELETS

Subtraction and integration help move us along the Simplification Slope. But in order to use either strategy, we must first accumulate a quantity of pieces, parts, and functions to then be sorted, discarded, and integrated. If we try to avoid the initial complexification activities and begin trimming or putting the pieces together too soon, we risk encountering what software programmers call premature optimization. The end result is less good than it might have been, because we overlooked important opportunities and approaches in our rush to simplify. Simplifying too soon is just as bad as complexifying for too long.

Edward de Bono puts it this way in his book *Simplicity*: "Sometimes a system starts off simple and then becomes more complex and then becomes simple again. This can be a normal process of evolution and adaptation to change. If the 'complex' phase is disallowed, then that system may be unable to evolve or adapt." Temporarily increasing complexity is therefore an important phase in the evolution of a design, because it increases our opportunity to find and identify the truly important design elements.

That is where the well-known KISS principle breaks down. It's the K in KISS that causes problems, because when we insist on "keeping" things simple and refuse to tolerate a temporary increase in complexity, we are focused on the wrong thing. We have placed simplicity ahead of goodness and are likely to end up with something that is simple but not very good. Perhaps a better guide would be to "Make It Simple Stupid," but that produces the unfortunate acronym MISS and I'm sure people would rather hit the mark than miss it. Maybe we should stick with KISS after all.

Whether we are KISS'ing or MISS'ing, the key to avoiding premature optimization is patience and diligence. It is important to first gather the necessary tools, talents, pieces, parts, and components . . . and only then can we apply them in the appropriate degree and trim out the extraneous.

Blogger Neil Mix made a related observation about the impact of simplicity. He calls it the Elegance Paradox:

> The design process is about whittling away distractions, making the obscure feel obvious, making the obvious feel implicit, and doing it without anyone noticing.
>
> To the untrained eye, your best work looks like you've done no work at all. If you've done a stellar job, then your design will feel utterly obvious.
>
> The Elegance Paradox is this: to create elegance requires entirely inelegant preparation, but nobody should be able to see that.

Good designers humbly move the product itself into the limelight, while their effort recedes into the background. This requires us to place the user's interests ahead of our own. Easier said than done, to be sure. We all want our effort to be recognized and appreciated. Unfortunately, the simplicity of an elegant design can obscure the blood, sweat, and tears that went into it, hiding the inelegant preparation phase shown in the shaded section of Figure 13. This is precisely why so many products are excessively complex—to make the designer's effort more visible—but such conspicuous complexity seldom serves the customer's interests. Keep in mind, most users do not care much how hard we worked. They are far more interested in how good the product is.

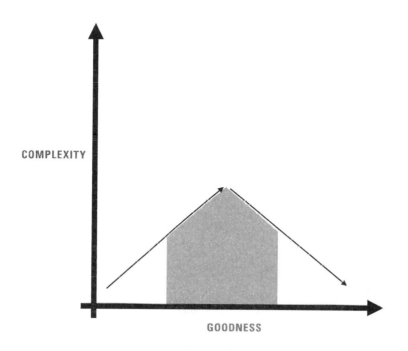

FIGURE 13: INELEGANT PREPARATION ZONE

When I was in college and the idea of a desktop computer was still relatively new, my classmates and I were given a homegrown word processer called Galahad. This program used a bewildering set of in-line codes and commands to produce super-high-tech effects like **bold** or *italics*.

Thanks to the wonders of the Internet, I was able to track

down a copy of the Galahad Users Guide. Chapter 2 (Papers, Letters, and Résumés) begins with the following codes:

```
.a lp=12 js=y bo=51 te=12 to=90 pn=1 nm=1
.st3,14,1
&dChapter%Two:%%Papers,%Letters,%Resumes .jc
&d[%Chapter%Two%] .hc
.t36 2—.zl
```

The rest of the guide continues in similar fashion. Looking back, I suspect the software was actually produced under a grant from whatever company provided the school's computer lab with printer paper, because one misplaced .a command would result in printing an illegible paper that had to be debugged, corrected, and reprinted. I'm saying I wasted a lot of paper back then. And the funny thing is, a properly formatted paper hid all those commands and gave no indication of how much work really went into it. Inelegant preparation indeed.

Naturally, some students became Galahad experts. Their skills were in great demand when it came time to write term papers, and debugging term papers became something of a sport, albeit not a fun one.

Galahad could do a lot of things, but nothing about it was easy or intuitive. You pretty much had to be an engineering student to use it. As I recall, even those of us who mastered Galahad weren't fond of it.

In contrast, modern word processors are intended for use by everyone, not just budding engineers. This is a good thing. The ability to turn a word bold at the click of a button makes things better for all users, including those of us who were previously willing and able to memorize complex formatting codes. Michael Dertouzos explained this phenomenon in his book *The Unfinished Revolution*, where he writes, "Whenever designers build utility for the least-skilled user, they enhance utility for all users." Providing for the "least-skilled" involves simplifying both the function and the interface, which frees up people's brain cells to perform higher functions now that we're not trying to remember whether js equals y and bo equals 51 or the other way around.

Randall Munroe draws the brilliantly simple Web comic *xkcd*, which is about "romance, sarcasm, math, and language." Comic #1306, titled *Sigil Cycle*, shows a repeating pattern of new computer languages that rely heavily on "weird symbols," followed by new languages that use more natural syntax. The former are intended for highly skilled users, while the latter are more accessible to a wider audience.

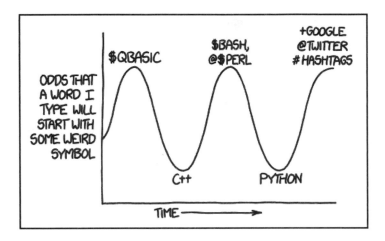

FIGURE 14: *XKCD* #1306: THE SIGIL CYCLE

The accompanying mouse-over text explains the cycle further:

> The cycle seems to be "we need these symbols to clarify
> what types of things we're referring to!" followed by
> "wait, it turns out words already do that."

This cycle persists because each new programming language aims to address the shortcomings of the one that came
before. When the previous language was simple and text-based,
the most obvious way to add new features and functions was

by introducing complex new symbols. But once we start using a complicated, weird syntax, the obvious improvement involves simplification. And so the cycle continues.

Ultimately, the design result we're aiming for is an elegant, graceful, streamlined solution. Such solutions are found in the bottom right quadrant of our graph, the Region of the Simple shown in Figure 15. In this area, we have integrated those metaphorical squares together into a single object, a cube. A cube is less complex than a collection of squares (it is one object, not six). It is also "more good" because we can do more with a 3-D object than a 2-D object. The secret is to put the pieces together in such a way as to produce something greater than the sum of the parts.

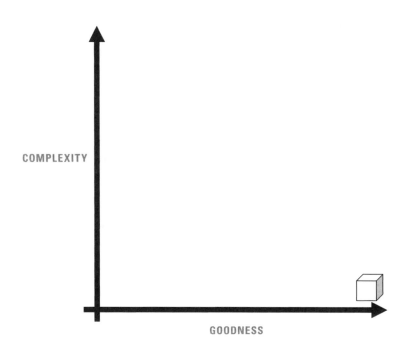

FIGURE 15: THE REGION OF THE SIMPLE

Note that simple thinking can be found in both the lower right and lower left corners. Because of this, profound simplicities may be mistaken for less valuable insights if the only thing we assess is the complexity of their expression.

So the challenge is to seek the simplicity on the other side of complexity, the elegance that results from experience and wisdom and is found in the lower right corner. We rightfully distrust the simplicity borne of naïveté and ignorance found in the lower left. But we do well to recognize that it is not the only place where simplicity can be found.

What happens next? Once we arrive at the lower right corner of the map, are we done? Is the journey over? Not exactly. Our design does not get to stay in the bottom right quadrant indefinitely, because the irresistible arrow of time exerts pressure to the left, in the direction of decreased goodness.

The object in question does not become more complex. It simply ceases to be as good as it was, because the goalposts have moved. The market changes. Technology changes. Needs and wants change. Yesterday's breakthrough becomes tomorrow's commodity.

As David Pye explained in his book *The Nature and Art of Workmanship*, "all designs for devices are in some degree failures," necessitating further development and improvement over time. The inevitability of failure is actually a good thing, because life would be pretty boring if goodness never decreased, if there was never a need or an opportunity to make something new and better.

It is important to understand that the time arrow is always present, pushing in the direction of decreased goodness. We feel the effects most when we try to stand still in the face of rapid change.

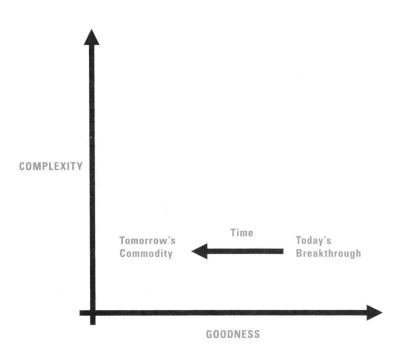

FIGURE 16: THE IMPACT OF TIME

In the world of consumer electronics, the arrow of time is very large indeed. Change happens fast and the iPod you got for Christmas begins to look like old news by April, as new and better products come on the market. The old device's goodness diminishes as the competition gets better.

This happens because an object's goodness is not entirely an inherent quality. To a large degree it is determined by comparison with alternatives. Last year's MP3 player may be good enough for our purposes, but it looks less good when placed alongside a device with better sound, larger memory, or a slicker interface.

It is worth noting that goodness is also subjective and personal. Certain retrophile hipsters have been known to happily listen to cassette tapes on a Walkman long into the twenty-first century even though iPods exist. But it would be difficult to make the case that a Walkman is significantly *better* than modern alternatives in any dimension other than a narrow definition of ironic fashion and a peculiar taste for old-fashioned approaches to musical recordings.

Because the arrow of time slides our design to the left, we eventually find ourselves once again needing to *increase* complexity to make the system better. Recall that the earlier transition from the Complexity Slope to the Simplification Slope meant we had to abandon behaviors when they ceased to be productive and helpful. We'll adopt a mirrored strategy here, as we return to practices we'd previously abandoned and begin once again to add pieces, parts, and functions to our design.

Figure 17 depicts the cyclical transitions between addition

and reduction, between making things more complex and less complex. It shows our initial progress along the Complexity Slope, which carried us to a peak in the center of the chart. At that point we began to simplify and move toward the lower right corner. The changing nature of needs and wants (as depicted in the aforementioned arrow of time) causes us to slide back toward the lower left corner, where the cycle begins again.

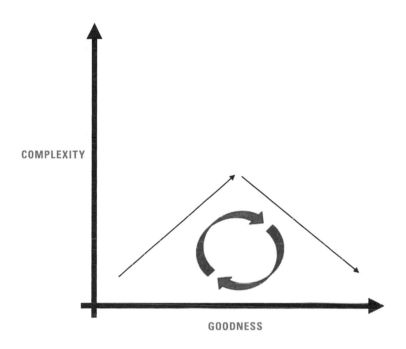

COMPLEXITY

GOODNESS

FIGURE 17: CYCLICAL TRANSITIONS

Let's talk about the word *cycle* for a moment. A cycle is a re-
curring sequence of events, and the Simplicity Cycle is indeed a
series of activities that we repeat when creating something. The
bedrock concept behind the cycle is simply that it is a cycle. As
designers, technologists, writers, or chefs, we begin our work
using one set of tools and pointing our efforts in a particular

direction, then we switch to a different set of tools and head in a new direction. Eventually, we return to the original tools and trajectory.

Of course, the idea isn't to exactly repeat the specific additions and complexifications we used the first time through, replacing pieces identical to those we just removed. We should be using the same strategies with *new* pieces.

Describing the Simplicity Cycle as both a cycle and a map produces a certain amount of intellectual tension. We expect a map to get us from Point A to Point B and to prevent us from driving around in a circle. Indeed, many of the illustrations and examples throughout this book describe singular journeys, or portions of journeys, where we have a starting point and a finish line and that is the end of the story. This piece-wise approach allows us to break down the cycle into components that are easy to grasp. And yet, the Simplicity Cycle is circular and repetitive rather than singular. If we are not careful the individual stories might cause us to miss the bigger picture. So bear in mind that each step, each phase, each pause is part of a larger cycle.

Time is the engine that drives this cycle. Without it, we could contentedly remain in the lower right corner, enjoying a perfect blend of simplicity and goodness. But the fact that time pushes our design toward the left, in the direction of decreased goodness, tells us something important about the value of simplicity. Namely, that simplicity is not static and all simplicity is not created equal.

Design master Don Norman explains in his book *Living with Complexity* that "simplicity by itself is not necessarily virtuous." He is quite right, and as we've already seen, there are many phases within the Simplicity Cycle in which a design might demonstrate such "unvirtuous" simplicity. In the initial stage, we encounter immature simplicity as the first parts of the design come together. Further along, we must take care to avoid premature optimization, which makes the design simpler without making it better. And finally, the elegant and desirable simplicity of a mature design can easily become unvirtuous if we fail to notice the inexorable effect of time.

Let's take a look at an example of technical simplicity that is practically virtue-free, not because time made it so but because of an error introduced by a different source.

My first day working in a new building was over. I was waiting for an elevator to take me downstairs when a ding sounded and a light indicated the car's arrival. As the doors opened, I paused, unsure whether to enter.

Much to my surprise, the pair of indicator lights located above the open door gave me no help at all in determining whether

I should enter the elevator, because instead of the typical top/bottom alignment, they were side to side. And instead of having arrows or points or labels, they were unadorned circles.

My fellow riders and I were left to guess whether the illuminated round red light on the right side meant the car was going up or going down. The alternative was a yellowish circle of light on the left, which helped not at all. I am of course familiar with the use of red and yellow to indicate Stop and Yield, but I am not aware of those colors ever being associated with any Up/Down conventions.

This was not a huge building and so it was not a huge deal, but a wrong guess meant I now had the opportunity to spend a little extra time getting to know the other people on the elevator as we went up instead of down.

This elevator signal was simple, but not virtuously so.

In a situation like this it is not hard to provide clarity without complexity. The lights could have been oriented one on top of the other instead of side by side. Problem solved. They could have been shaped like arrows or triangles instead of circles. Problem solved again, with no increase in complexity.

Can't change the orientation or the shape? Okay. Just paint arrows, triangles, or the letters *U* and *D* on the circles. Can't paint on the circles themselves? Fine—put an indicator *next to* each circle. Can't do that? Aw, come on now . . .

The point is, the elevator people had options. They picked the worst one.

An indicator that doesn't indicate anything is pointless. In fact, it's worse than pointless, because it creates the impression of communicating something even though the signal is actually content-free.

Was that indicator simple? Yes. But that's nothing to brag about.

The elevator story is true and really happened to me, but one part might be misleading, namely the subtle implication that someone designed the lights this way. I am quite sure nobody would ever design such a thing and I don't want to give the impression that these lights were the result of anyone's conscious choice. Most likely what we have here is a failure to coordinate.

Here's what I think happened: Somebody designed a top/bottom light set. Someone else built a ceiling too low to accommodate that orientation. Someone else purchased the light set for use in that hallway. Someone else installed the set sideways. Someone else signed off on the job. Then I showed up and felt confused.

That's a lot of someone elses—and I probably overlooked a few. Recall that "a lot" of something indicates a high level of complexity.

Is the light badly designed? Maybe. Or maybe it is simply the wrong light for this hallway, selected by someone far removed in space and time from the location's physical reality and un-aware of its limitations and specific requirements, and installed by someone who did not actually have to ride the elevator. This means that light is almost certainly an example of how a group's complexity makes coordination and communication difficult. It is a failure of organizational design rather than physical design.

Simpler, smaller, more coordinated teams don't usually have that kind of problem, but if they do, they are close enough to the thing to recognize it and remedy it. Members of a small team are more likely to proactively add some kind of label to the lights. Members of a big team, in contrast, either expect someone else to do it or do not feel empowered to make the correction.

Of course, complexity directly affects our technologies, too. Software locks up when it receives conflicting inputs, can't quite decide what to do, and gets stuck in a loop. This is more likely to be designer error than user error. Elegantly simple computer programs don't have this problem.

But as the elevator indicator showed, ambiguous simplicity can be just as problematic as ambiguous complexity. Removing information-rich indicators in the name of simplicity does not make the thing better. It makes the thing worse. True clarity is the result of thoughtful simplicity, not unnecessary complexity or superficial simplisticness.

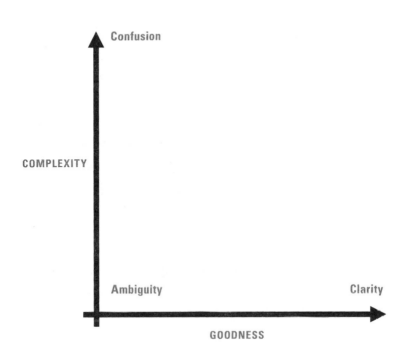

FIGURE 18: CONFUSION, AMBIGUITY, AND CLARITY

If at any time in the previous chapter you felt that things were getting too complicated, take heart! You are correct that things got a bit complicated, and that's not an accident. It's all part of the process. Fortunately, those momentary complexifications are behind us and we can now move into a simpler, cleaner description.

Giving each arrow and area a name helped present the concepts and talk about the arrows and areas themselves. The complexity of all those vocabulary words and specialized jargon was necessary and useful . . . and temporary. Now the foundation is in place and we are ready to use the diagram, so the labels are largely unneeded. We can strip them away and instead focus on the concepts themselves, simplifying things considerably.

Laying out the Simplicity Cycle diagram is a starting point, aimed to familiarize us with a tool for future actions. We are done assembling the pieces, which means we're ready for the next phase, where we actually put the map to use. No doubt you have some ideas about what you'll do with the diagram and where you'd like to go. In the next chapter we'll take a look at several ways to do just that.

CHAPTER 4

Using the Simplicity Cycle

Like any book, this one owes its existence to a large group of people. Engineers and professors, agents and editors, family and friends all contributed to bring these pages to life, not to mention the legion of speakers, thinkers, and writers who shaped my understanding of engineering, design, and life in general. But as I reflect on how this book came to life, the person who really sold it just might be the dentist.

I'd been doing presentations and seminars based on the Simplicity Cycle for several years when I met her. My audience members are usually technology types—aircraft designers, computer programmers, engineers of all stripes. On the list of occupations I did not have in mind when I came up with this concept, dentistry is certainly in the top ten. Not that I have anything against dentists (okay, maybe a little), I just never imagined my work would have any relevance for them.

Nevertheless, after I shared the Cycle with a group of young professionals one afternoon, there she stood, with her perfectly brushed and flossed teeth, telling me how closely her experience at a dental clinic lined up with the pattern I'd just described. The clinic's procedures and policies, as well as their tools and technologies, all too often were unnecessarily complicated in ways that made things harder, slower, less effective, and more expensive. She shared several ideas for specific simplifying improvements she wanted to try, now that she was thinking in terms of goodness and not expecting complexity to automatically make things better.

My new friend the dentist was not the first person to connect the Cycle with an application outside the realm of high-tech. That would be me. I already knew from firsthand experience that it describes the writing process pretty well. And in the very early days as I was hashing out the concept, an artist whose work I greatly admire said, "Dan, you just described my art-making process." Then again, that guy is a fairly technical artist and what he does is not all that different than what some of the aeronautical engineers I've consulted with do. So, the fact that this idea resonated with writers and artists was not surprising. But a dentist? Man, I did not see that coming.

Her response motivated me to seek out wider opportunities for the Simplicity Cycle, beyond the technical realm. Her response also points to an important question, perhaps the crux of the whole book: how can we use the Simplicity Cycle? The dentist used it to identify areas in her practice that could use some improvement, but what about the rest of us? What can we do with it?

I find the map metaphor helpful in answering that question. The Simplicity Cycle is a map that can help us figure out where we want to go and can help us find the paths that will take us there. It enables us to think in new dimensions and discourages us from simplistically equating complexity with value.

Whether we are designing a business practice, a dental procedure, or an art installation, this little map shows that some routes will take us toward making things better, while other routes point us in the opposite direction. So the map is a useful guide for individual behaviors and decisions as we design and create things. But the real power comes to light when we use it with other people.

When teams get together to work on a design, it is sometimes difficult to verbally express concepts related to complexity, in part because we English speakers somehow ended up with a single word (*simple*) that means both "easy" and "uncomplicated," as if those two concepts were synonymous. To make matters worse, we also use the word *simple* to describe someone with limited mental abilities. No wonder there is so much complexity out there. Our language itself nudges us away from simplicity by befuddling and devaluing its meaning.

To see why this is a problem, let's try a little challenge. Go get a mug of superhot coffee and balance it on your nose. [*Ed. note—this challenge is provided for illustrative purposes only. Please do not attempt.*] This is not a complicated task. There are only two elements, the mug and the nose, although if we include gravity, I suppose there are three. But despite not being complicated, it is not easy, is it? How many of you burned your faces just now while trying to get the balance just right? [*Ed. note—hopefully none of you.*]

The Coffee-Face Experiment shows that something can be simple without being easy. It also illustrates how failing to make that distinction can lead to all sorts of problems.

The Simplicity Cycle aims to help us resolve this situation without rewriting the dictionary by providing a *visual* vocabulary that augments our verbal capacity to describe and discuss designs. The various slopes and regions in this chart allow us to distinguish between a thing that is simple (easy) and a thing that is simple (uncomplicated) in a way that mere words often cannot. Most critically, it forces us to assess whether or not the thing is any good. As you may recall from an earlier chapter, that's really the point.

How do we do this, exactly? When I'm working with a design team, I like to draw a blank Complexity-Goodness graph on a whiteboard, then ask my teammates to indicate where they think the current design resides and where we should head next. That's a simple approach to an important task. Whether or not it's easy depends on who is in the room and what sort of design they're working on.

Easy or not, this approach leads to lively, insightful discussions analyzing the current design and exploring specific ideas about how to improve it. Introducing this new communication mode minimizes the time spent talking past each other, as when one person insists that a design is simple while another argues that no, the design is actually simple but not simple, while a third person says it needs more simplicity and everyone else in the room just eats the donuts.

Instead of arguing whether or not simplicity is simple, or whether any sense of that word applies to our design, the Cycle diagram instead leads us to talk about how good our design is, how the various elements of the design contribute to (or decrease) goodness, and what steps can be taken to move in the direction of greater goodness. No longer can we contentedly assume each addition should be embraced and included. Instead, when someone proposes adding a new mode, new feature, new step, or new component, the team must answer the question of whether or not the addition is an improvement.

Along with helping teams to evaluate the wisdom of additive methods, the Cycle diagram often prompts design teams to pursue improvement strategies they might have otherwise overlooked. Reductive approaches such as subtraction and integration become more visible and attractive, opening up alternative approaches that can lead to powerful breakthroughs.

No whiteboard? No problem. The same approach is just as effective on a piece of paper, a napkin, or the back of an envelope. Maybe even more so.

Introducing a visual vocabulary like this to a group can be a major boost to communication and problem solving, largely because these visuals have less baggage than words like *simple*. We approach the lines and arrows and swirls with fresh eyes and explicit meanings, free from unconscious assumptions. Drawing takes us away from our default way of talking and leads us to be more precise, thoughtful, and deliberate, without slowing things down. Even when the Simplicity Cycle becomes a familiar tool in the group's toolbox, it is still sufficiently different from the more typical verbal communication to free us from status quo thinking. Plus, over time it can change the words we use and make them more useful.

Along with functioning as a visual vocabulary, the Simplicity Cycle is also a kinetic vocabulary. As we stand before the whiteboard, marker in hand, we must move our bodies to express ourselves rather than relying solely on words. We do not merely say, "The design is too complicated." We say, "The design is *here* and needs to go *there*," as our hand sweeps across the board to draw a line or a curve. This awakens an entirely different region of the brain and unleashes previously untapped imaginative resources.

In his book *The Reflective Practitioner*, Donald Schön describes this sort of thing as a "drawing/talking language." He analyzes a vignette about a pair of architects collaborating on a design and highlights the way they "draw and talk their moves in spatial-action language." He points out that "drawing reveals qualities and relations unimagined beforehand," and shines a light on places where words alone cannot reach.

Using the Simplicity Cycle as a "spatial-action language" not only changes how we transmit and receive information; it also changes how we produce it. We think differently about things when we express thoughts in the form of movement. Schön explains that drawing is a type of experiment, a way to test theories and make discoveries. The person holding the pencil does not always know what the result will be until after the lines have been drawn. Drawing is thus a form of thinking out loud and, when done with a group, is a way to collaboratively explore multiple alternatives that might otherwise be overlooked.

As we have these conversations, either with a group or in our own heads, the Simplicity Cycle map confronts us with many questions. One of the most common is this: "How can I tell where I am?"

Well, if we're starting with a blank sheet of paper, we're in the bottom left of the chart and we probably know it. Similarly, when we arrive at the Region of the Simple, we'll know. Trust me, we'll know. So all that's left is to determine where in the central cloud we are.

That may be a common question, but *it turns out that's probably not the right question.*

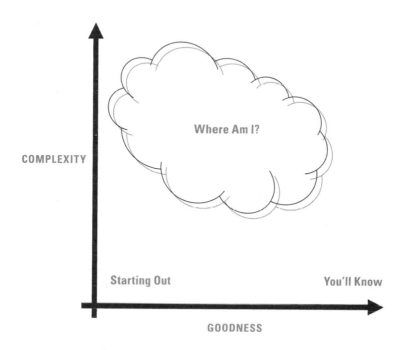

FIGURE 19: WHERE AM I?

Our specific location in the "Where Am I?" cloud at any moment in time is less important than our trajectory. Rather than asking, "Where am I?" we should focus on figuring out, "Where am I going?" In engineering terms, our vector matters more than our position and the important thing to identify is whether the latest change to the design makes it better or worse.

One reason our location does not matter much is that the prescribed behaviors for the upper left corner are essentially identical to the prescribed behaviors for the center of the chart. In both cases, improvements require integration, simplification, and streamlining. So don't spend too much time trying to narrow down a precise location. Think instead in terms of where we're heading.

Keep in mind that when we talk about the design process, we're largely talking about making changes to the design. We add things or subtract them, making our object simpler or more complex. So the question is whether the transition from version 1.3 to version 1.4 represents an improvement or merely a complexification. Did our trajectory carry us toward the right, in the direction of increased goodness, or toward the left, where goodness decreases?

The answers to these questions help us identify what type of design behaviors we should adopt next. If our previous step made things more complicated and worse, our next step should probably involve simplification. But if the previous step made things better, we may want to continue along that former path.

Now, if there is one point we could try to identify, it's the "peak of complexity," the point of critical mass where increases in complexity reduce goodness. How can we tell if we're there?

The short answer: we can't. Cliff Crego explains:

Where the climax of complexity comes we can never know for sure, but natural movement always begins and ends with simplicity.

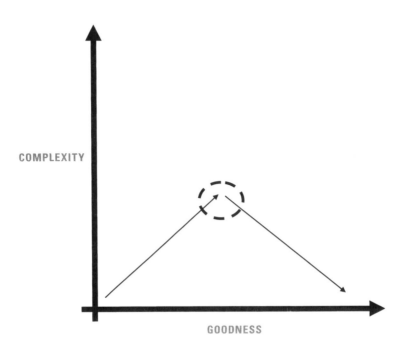

FIGURE 20: THE CLIMAX OF COMPLEXITY

The slightly longer answer is that experience, intuition, the insight of others, and hindsight are the best ways to make the assessment. The more times we go through this experience, the more likely we are to recognize the pattern when it is repeated. Sometimes we just have a gut feeling that we've hit the peak and it's time to shift our approach. Alternatively, people with fresh eyes can see things we can't. And looking back, we can almost always see the point where our design hit the inflection point and started to get worse because we continued to add complexity unnecessarily, or the point where we changed our design approach and began a course of simplifications that dramatically improved the situation.

Reflecting on our design efforts (both in the moment and afterward) helps translate experience into expertise, turning what we did into what we know. This also bolsters our intuitive capacity for recognizing when we have hit the peak and should begin the transition from genesis to synthesis. The benefit of this practice is not merely for our current design but also for future endeavors. Spending some time thinking about today's project helps improve our prospects for tomorrow's.

Recall the earlier story about Microsoft's Vista and Windows 7 software. The Vista team continued adding new features (to include fifteen shutdown modes) long after passing the peak of complexity. The final product left much to be desired. Not wishing to repeat this failure, Microsoft adopted a new approach that was more sensitive to the impact of complexity and provided incentives to software coders who found ways to streamline and simplify the next product.

Previous comments notwithstanding, there is not necessarily "one" critical mass point we should aim for. As Figure 21 shows, it is more of a region than a point.

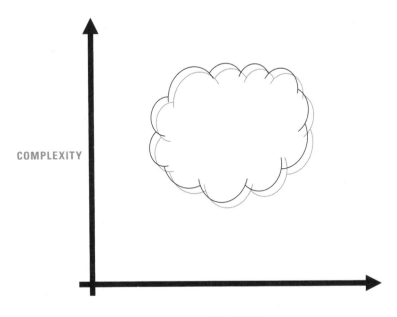

FIGURE 21: CRITICAL MASS

Design conversations should not be limited to designers. At some point, we may also want to talk about the project with people outside the design team, such as bosses, customers, or investors. The Simplicity Cycle can be useful here, too.

Many consumers automatically assume a digital camera with 100 photo modes is superior to a camera that only has 84, because complexity is so easily mistaken for sophistication. Never mind that we only ever use one or two of those modes—people still prefer to pay a little extra for all that other stuff we neither need nor understand. But imagine if we could help consumers be more thoughtful, more deliberate, choosing instead to buy things that are a better fit for their needs. This approach serves the consumers interests *and* improves the quality and reliability of the product, because simpler designs have fewer ways they can break. Fewer problems equals happier buyers. Everybody wins.

I'm not suggesting product designers have to gather all their potential consumers around a whiteboard and walk them through the Simplicity Cycle diagram, although that would be awesome. Instead, I'm suggesting that a deliberate emphasis on simplicity can and should be a selling point. The key is to make the connection between simplicity and goodness, rejecting the usual tendency to play up unnecessary complexity as a desirable feature. Some companies like Apple do this well already. Others have lots of room for improvement.

This shift in preference from complexity to simplicity is neither automatic nor intuitively obvious, so we'll need to make some effort to bring people along and demonstrate the merits of this approach. It's something to do *with* people, not *to* people, and it starts with a conversation. Depending on who we're talking with, that conversation may not involve sketching x-y axes on a whiteboard, but once we are familiar with the Cycle, we can find many ways to express it, perhaps through stories and examples that are specific to our situation.

The point is, there are many ways to use the Simplicity Cycle—as a design prompt for a solo project, as a communication enabler and creativity prompt for a team, and as a way to discuss a project's value with the boss or with customers. Like my friend the dentist, you'll probably find ways to put it to use that I never imagined, whatever type of work you do.

In the next chapter, we'll use the diagram as a lens to look at the design of work itself. We'll focus on the relationship between effort and achievement and spend some time looking at how complexity affects the way we relate and communicate.

CHAPTER 5

On Hard Work and Design

Imagine it's a warm spring day. The sun is shining, the birds are chirping, the scent of lilacs fills the air. We would like to open a window in order to more fully appreciate nature's beauty, so we grip the windowsill and press upward.

It remains shut. Darn it.

We push harder, but the window refuses to cooperate. What is our next move? We could get a tool—a crowbar, a screwdriver, a hammer—anything to convey an increased amount of force to the window. If we're interested in a permanent opening in the wall, dynamite would certainly produce the desired effect although the lilacs outside would undoubtedly suffer. Or we could embark on a six-month weightlifting regimen, to increase our arm strength in order to overcome the window's resistance, a strategy with a bonus health benefit. Then again, by the time we're sufficiently bulked up, winter will have arrived and the lilacs will have wilted, so this approach is unlikely to be substantially more satisfying than the dynamite.

Fortunately, there is another option. We could use a single finger to gently swing the window latch into the unlocked position.

Ah, that's better.

Trying really hard without making progress often indicates something is wrong with our approach or our understanding of the problem. There may be an unseen barrier involved, and removing the barrier may be a simple matter of stepping back and considering the situation from a different perspective. Overcoming obstacles often requires more thoughtful effort and less brute force, more simplicity and less complexity. Maybe we need to put down the sledgehammer and unlatch the window.

How often does this happen in our lives, both professionally and personally? We push hard to get promoted, to get a date, or to create our magnum opus, but instead get passed over, turned down, or otherwise underwhelmed with the results. Should we have tried harder? Or was trying harder part of the problem? Maybe we were pushing in the wrong direction, adding complexity and effort where simplicity and ease were called for. Maybe a pause or a trim would have led to a better outcome.

This is not to denigrate hard work, but let's not fetishize it, either. When we are in the zone and operating at our peak ability, things move smoothly and that's okay. In such situations, progress may appear and even feel easy. This does not mean it isn't real progress. That does not mean it isn't good.

In his autobiography, filmmaker Akira Kurosawa bemoans the ways "human nature wants to place value on things in direct proportion to the amount of labor that went into making them." He argues that a thing's value ought to be measured in terms other than how hard it was to do. I think he's on to something.

To be sure, there are times when work feels like work. This, too, is okay. Sweating and straining are not always things to be avoided, and if we want to run a marathon, we have to put in the miles. However, our marathon training should include learning to run with good form, to provide maximal return on our exertion, and in a sense to minimize the amount of effort required. Not only does good form make for an efficient runner; it also reduces the likelihood of injury.

The question of effort and ease matters because when we place value on things based primarily on how difficult they were to accomplish, we head toward the upper left corner of our map, toward greater complexity and diminished goodness. Things in that area are very difficult indeed and if we don't know any better, that's where we'll head because we've mistaken difficulty for goodness. A more thoughtful approach sees that difficulty in itself is nothing to brag about, that signs of strain are signs of opportunity for improvement rather than signs that all is well. When something is difficult to do, we might be facing a situation ripe for simplification, even if nobody else sees it. Particularly if nobody else sees it.

For generations, travelers carried heavy luggage through train stations and airports. Sometimes, clever travelers would pay a few coins to rent a wheeled cart upon which to stack their bags, but the idea of putting wheels directly onto the bags themselves didn't really take off until 1987.

When wheelie bags finally appeared on the market, some macho types scoffed at the implied weakness of people who can't carry a bag, while countless others around the world, including esteemed experts in the prestigious discipline of luggage design, surely uttered the phrase "Why didn't I think of that?" The answer is that they treated the strain of carrying a suitcase as inevitable. That burning sensation in your shoulder as you lug your luggage from place to place was just one of the unavoidable consequences of travel.

In hindsight, putting wheels on bags looks like an obvious solution, but people who probably should have known better failed to see it, largely because they failed to recognize a remarkably obvious problem. Probably they needed a nap. As someone who's old enough to remember carrying a suitcase through an airport or two, I know I needed one.

The funny thing is, the wheel is a fairly old type of technology. The wheeled luggage solution therefore did not require any major scientific breakthrough before it could be implemented. The wheel is also relatively inexpensive, so it could be added to a bag without significantly increasing the cost. And the wheel is simple, with no fragile electronics, no need for batteries or computer chips. Where's the downside? Or rather, what took us so long?

It is such a basic concept that we all might be tempted to kick ourselves for not coming up with it. But do not be fooled. If the solution was truly obvious, people would have thought of it already and figured out a way to implement it. Making this type of improvement requires a very specific mindset, one that views difficulty as a signpost that change is needed and possible, rather than as an unavoidable aspect of the human condition. It requires us to seek out the clever simplicity on the other side of complexity, to ask questions other people are not asking. And as generations of suitcase-carrying travelers can attest, that mindset is rather rare.

Fortunately, it is possible to cultivate it and deliberately point our minds in that direction, by paying attention to things that are difficult and looking for opportunities to add a wheel or move a window latch. The Simplicity Cycle highlights that particular path and that mode of thinking, by pointing out that the path to increased goodness often involves removing unnecessary effort rather than tolerating it.

The Simplicity Cycle reminds us that goodness matters more than difficulty or complexity. It also reminds us that solutions that are simultaneously simple and good can exist. Maybe not all the time, but certainly some of the time. By mapping out the location of these potential solutions and showing some of the paths that will take us there, the diagram opens up new vistas and opportunities to explore. It points us in the direction of a different type of skill and mastery, one that accepts difficulty where necessary but does not automatically equate strain with progress.

One of the problems with excessive strain is that it leads to exhaustion, and when we're tired, everything is harder. Not because the world becomes any different when we are wiped out, but because fatigue makes us stupid. It clouds our judgment and tempts us to needlessly adopt brute-force methods, trying to solve problems with the maximum amount of effort and minimal amount of cleverness. When we are tired, we are more likely to pick up the sledgehammer or dynamite, overlooking the window latch.

If you ever find yourself in a whirl of complexity, straining without making headway and pouring out resources with no visible return, it probably makes sense to take that pause we talked about earlier. Sometimes, a nap can make all the difference in the world.

FIGURE 22: SLEEP ON IT

Rest is important but anyone who tries to nap their way to excellence is bound to be disappointed. Effort matters, and in his book *Outliers*, Malcolm Gladwell famously wrote that mastery of any skill requires ten thousand hours of practice. He admits that is a suspiciously round number, but his point is still worth considering. When we approach it correctly, practice does indeed make perfect and such perfection does not happen overnight. It takes upwards of ten thousand hours' worth of effort.

Practice is about effort, yes, but practice is also about making and fixing mistakes, identifying and removing friction from our efforts. It does not entail merely doing the same thing over and over, but instead involves a thousand small tweaks as we experiment with new techniques and modifications of former ones. Eventually we are able to focus on the techniques that work, honing our ability to perform. In doing so, we move our practice from the center of the Simplicity Cycle to the lower right corner. Toward perfection, however fleeting that perfection might be.

I wonder, does a pianist play her instrument in order to keep her fingers nimble, or is it the other way around? Does she not cherish nimbleness because it produces music? In the same way, is not simplicity admirable and virtuous because of what it produces, what it allows us to do? Do we not simplify in order to achieve higher goals, rather than simplifying for the sake of simplicity itself?

It is worth restating: Simplicity is not the point. It is a tool, a discipline, a practice that helps us do other things. Things that matter infinitely more. Simplicity is an essential ingredient in a well-designed product and a well-lived life, but it is only an ingredient.

Nevertheless, simplicity is an important ingredient. When we leave it out, the result can devolve into an expensive, overengineered monument to difficulty (for the designer and consumer alike).

We say a thing is overengineered when its performance far exceeds its environmental demands. Such a device uses gold plating when aluminum will do, operates at frequencies beyond a human's ability to appreciate or speeds beyond a human's ability to operate. It might have a designed life span far beyond the underlying technology's expected relevance. Overengineered designs are extremely vulnerable to disruptive innovations that provide less capability but are better aligned with market needs, as Clayton Christensen explained so brilliantly in *The Innovator's Dilemma*. Providing consumers with high-end products might create a barrier to entry for competitors who cannot provide comparable quality, but it also creates an opening for a low-end, good-enough entry into the market that just might prove popular with high-end clients.

Imagine if someone in the 1980s had designed a Walkman robust enough to last one hundred years. The accompanying cost would be high, in large part because of the complexity and redundancy necessary to achieve such objectives. Overengineered products are generously described as "high-end," when a more accurate description may simply be "pricey."

A better alternative is the highly engineered product, which is sleek, elegant, and focused. It exhibits alignment between what it is designed to do and what it actually has to do. Clearly, we have options and alternatives when we design things. Overengineering is not a good option.

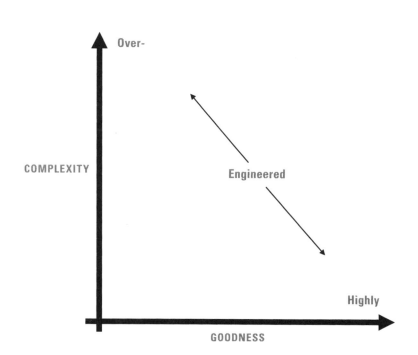

FIGURE 23: OVERENGINEERED/HIGHLY ENGINEERED

Speaking of options, the opening line of Shunryu Suzuki's best-selling book *Zen Mind, Beginner's Mind* says, "In the beginner's mind, there are many possibilities. In the expert's mind, there are few." Don't read past those sentences too quickly. Assume the lotus position and take a moment to ponder them before moving to the next paragraph.

Now, I'm not a Zen master, not by a long shot. I'm a garden-variety Methodist and when it comes to Zen, I'm strictly an amateur and dilettante who may have no idea what he's talking about. Does this mean I have beginner's mind (*shoshin*)? Maybe, but how would I know?

Given my almost complete ignorance of the subject, I won't pretend to understand *shoshin* in any sort of enlightened manner. However, I would like to offer a few observations, the first of which is this: Suzuki is making an observation.

Let's reread his words: In the beginner's mind, there are many possibilities. In the expert's mind, there are few. In those two sentences, Suzuki is neither complimenting nor criticizing anything. He is *noticing*. His comments are descriptive, not prescriptive. He is pointing out a difference between experts and beginners, not passing judgment or comparing relative values. Thus, any assumption that one type of mind is better than the other is something we bring to the text, not something we find there. Like the ancient story of the Zen master who continued pouring tea into the student's overflowing cup, we need to empty ourselves of our preconceptions before we can come into contact with reality.

If we take Suzuki's statement as a straightforward indictment of narrowness within expertise and a blanket praise of beginner's imaginative powers, we not only misunderstand him, we also misunderstand Zen itself. And if we persevere long enough to get to page 2 of his book, we will hear Suzuki-roshi tell us, "For Zen students the most important thing is not to be dualistic." Thus, his dual framework of beginner and expert is not what it might appear to be at first blush.

Yes, beginner's mind is a valued practice in Zen, and many practitioners—including Suzuki—talk about cultivating it and "always embracing *shoshin*." This might seem to imply that seeing more is the same as seeing better. But let us bear in mind that Zen also values restraint and stands in stark contrast to the common Western philosophy of "more is better." So, yes, beginner's mind sees many possibilities and expert mind sees few. Who sees better? Make the sound of one hand clapping and then I'll tell you.

Suzuki's observation about beginners and experts illuminates several important aspects of the Simplicity Cycle. When we first set out on a new project, few decisions have been made and many options are still on the table. Faced with a lot of options, the beginner adds and collects and creates, moving from the lower left corner to the middle of the map. This expansive, inclusive worldview is important and productive as long as it sets a foundation for subsequent simplifications and is not allowed to carry the design up and to the left, in the direction of foolish complexity.

In this early phase of a design, the proverbial Cone of

Uncertainty stretches out wide before us and holds "many possibilities." Then, as we make decisions and advance the design, we narrow down our future options until we have "few."

At that point, the student must become the master and develop an ability to distinguish between possibilities, identifying which are viable and which are not. This culling of possibilities can be described as "seeing fewer," as the unproductive and unwise possibilities are cast aside. In this way, the designer now pursues simplifications and a narrower set of options, transitioning to a new trajectory that carries them down and to the right.

But just as the beginner must take care to not allow their expansive vision to carry them to a place where complexity overwhelms goodness, the Master must take care to not let experience lead to a foolish narrowness where they see less well than they once did and overlook possibilities that should be explored or embraced. The Master must be on guard against the tendency to follow only the comfortable, proven paths and thus overlook the Right Path, which may be more visible to a novice.

While a beginner facing infinite possibility may find it impossible to take the first step, Masters face a different danger, that of arrogant narrowness and blindness to new alternatives, making it difficult for them to take a step they have not taken previously. Humans naturally develop intellectual inertia, latching on to certain ideas and beliefs to the exclusion of others. Even an enlightened one can fall prey to unwise habits of thought and mistake them for inevitable truths.

Thus Zen practitioners at all levels of experience are encouraged to cultivate *shoshin* and maintain an openness to surprising possibilities. In order to do this, it helps to have several beginners around, to help remind us of exactly what a beginner's mind looks like. Perhaps this is one of the reasons that beginners and masters are often found in proximity to each other, a situation that benefits both.

For example, a beginner facing a million options may be paralyzed into inaction because he or she lacks the ability to distinguish and judge between alternatives. That is where the Master's narrowness comes in, to break the beginner free from immobilizing uncertainty by pointing them in the right direction and helping them take their first steps down the Noble Eightfold Path.

One of my favorite Zen stories illustrates this nicely. It tells of a novice monk who asked Master Zhaozhou to teach him. The apprentice saw so many possibilities he was unsure which one to pick or what to do. Zhaozhou, in whose mind there were few possibilities, simply asked if the monk had eaten, and when the monk answered in the affirmative, the Master told him to go wash his bowl. Upon hearing this, the novice went away enlightened. The story doesn't mention it, but I suspect Zhaozhou went away enlightened as well.

We conclude this discussion of Zen the same way we began, with Suzuki's observation: In the beginner's mind, there are many possibilities. In the expert's mind, there are few. As our design efforts carry us through the Simplicity Cycle, bear this koan in mind. As complexity rises and falls, as goodness increases and diminishes, as possibilities open and close, bear this koan in mind. As we expand our designs through the inelegant preparations that eventually become invisible or as we trim our designs into elegant simplicity, bear this koan in mind. Whether you are a novice or a master, a beginner or an expert, bear this koan in mind. And don't forget to wash your bowl.

We turn now from ancient Zen wisdom to modern techno foolishness. I'm talking, of course, about Facebook. In this next section, we're going to shift from looking at a specific master/novice relationship and instead take a look at the role complexity and simplicity play in other types of relationships, beginning with an observation from everyone's favorite social media platform.

Changing one's Facebook relationship status to "it's complicated" is seldom a sign that all is well. The transition from a simple, familiar, stable state such as "in a relationship" or "married" to the ominously ambiguous "it's complicated" indicates a state of flux and uncertainty that few people deliberately seek out and even fewer enjoy. As a brief, temporary phase, "it's compli-cated" can be an important step that ultimately leads to a more satisfying resolution, but we pity the poor souls who perpetually reside in this limbo.

The idea that relational complications are a sign of trouble does not only apply to our private connections. Excessive inter-personal complexity can make our professional dealings difficult as well. This is no great insight, but it might be worthwhile to spend a few minutes considering strategies that help us move toward relationships that are both simpler and better, because design is fundamentally social. Design involves working with people—colleagues, bosses, customers, suppliers. It involves conversations and negotiations, compromises and concessions. And just as unnecessary complexity in a design can weigh it down and make it less good, unnecessary complexity in our rela-tionships can lead to all sorts of problems.

The first step is to figure out what type of organizational complexity we're dealing with. Three common sources of com-plexity are size, structure, and specificity. We'll look at each one in turn, starting with size.

Large organizations are intrinsically more complex than small ones because they have more parts—more people, more divisions, more organizational interfaces. This increase in com-plexity feeds on itself, because hiring ten new people requires us to hire an eleventh to oversee them all, which probably creates an additional layer in the organization's structure, and things just get crazy from there. We might debate whether or not the increase in size makes the organization better (sometimes it does, sometimes it doesn't), but there is no denying bigness makes things more complicated.

And then there is the question of structure. Regardless of size, certain organizational forms are inherently more complex than others. For example, in matrixed organizations, lines of responsibility are shown on organization charts via dotted lines, solid lines, and dashed lines of various thicknesses and colors, which change depending on the day of the week and the phase of the moon. Specialized software is sometimes required in order to depict the complexities of the matrix, where control and accountability follow different paths, and where influence and authority are divided from each other.

As with large organizations, a complex organizational structure conveys certain benefits and strengths, enabling the organization to do things it might otherwise be unable to accomplish. Do the benefits of this approach outweigh the costs? That's the question to ask, isn't it? As we've seen several times throughout the book, complexity can be valuable and essential. The key is to know when to stop and to make the effort to simplify when things get out of hand.

The third source of complexity, specificity, has less to do with the external structure of the organization itself and more to do with how people approach the work. A formal, strictly defined approach to work, relying on huge binders of policy and intricate process diagrams, may be established for the purpose of simplifying things by always having an answer to the question "What should we do?" However, in practice the result is more complexity, not less.

A big specific rulebook is more complicated than an informal set of guidelines and principles, because when we have a large collection of rules that aim to control and guide every aspect of operations, we must also deal with interactions between the rules, addressing the inevitable areas where rules come into conflict with each other. And then there is the question of areas the rules do not address, or address ambiguously, which requires a whole new layer of rule making and interpretation. To make matters worse, this formal approach tends to foster a parallel discipline of loopholery, where specialized experts add their own layers and twists to help people interpret the rules in a variety of ways. Simpler? I don't think so.

Large and complex organizations tend to gravitate toward this formal, legalistic approach to work, creating a perfect storm of size, structure, and specificity where all three sources of complexity amplify each other. But even a small organization with a straightforward structure can find itself in complicated territory by embracing a strictly specified mode. As with size and structure, specificity conveys several benefits and can contribute to an organization's performance. It can genuinely make things better. But it is important to recognize that specificity also makes things more complex, and as with any type of complexity, there is a cost associated.

Size, structure, and specificity are not the only sources of complexity in an organization's relationships. Sometimes it's all about the drama. When the person in charge of testing a new piece of equipment threatens to cancel the entire test program every time there is a hiccup in the paperwork, or when the security expert helpfully offers to rescind previously granted approvals because of a minor change in a seating arrangement, or when the legal counsel confidently predicts that the smallest divergence from his or her recommendations will certainly result in everyone going to jail for life, the complexities being introduced are more the result of personalities than anything else. This willingness to put the nuclear option on a hair trigger introduces a lot of friction and unnecessary complexity as the rest of the team walks on eggshells lest our drama-loving partners start running around like chickens with their hair on fire. Two words: don't be that guy.

But some people add drama (and therefore complexity)

whether they want to or not, by virtue of their position more than their personality. I'm talking about bosses and other people in positions of authority. See, bosses are dramatic because extremes are dramatic, and the top of the pyramid is an extreme position. Even when they communicate in calm and soothing tones, a boss's presence alone is enough to increase everyone's level of dramatic tension. This is not necessarily a bad thing, and sometimes it's precisely what the situation calls for. But bosses would do well to maintain awareness that their presence and involvement raises the stakes considerably and can complexify things in unproductive ways.

Okay, so organizational complexity comes from many places, including several not mentioned here. The point is that if we first spend a little time identifying the primary contributors, it's not too hard to find strategies for dealing with them.

If the boss's presence makes things complicated, then he or she might need to step away for a bit. Sometimes the best thing a boss can do is be elsewhere.

Is the organization's size dragging things down? Get smaller. This does not necessarily mean downsizing. Instead, it may be a matter of forming smaller subteams within the larger units. Large churches, for example, often deal with the challenges associated with growth by building and nurturing small groups that are able to provide fellowship and support that a large group simply cannot. These self-sustaining cells are still connected with the larger congregation, but they reduce some of the complexity-related pressures on the organization as a whole.

Is the complex structure making it harder to get work done instead of easier? Simplify the organization by making relationships more direct and less vague, more solid line and less dotted line. If we can't actually redraw the matrix into a simpler structure, we can still take steps to reduce ambiguity and confusion, such as having face-to-face discussions with actual human beings, rather than communicating via email and other distancing methods.

Is specificity getting in the way of progress? Take a page from Brazilian businessman Ricardo Semler. In his book *The Seven-Day Weekend,* Semler talks about how his company Semco dealt with unproductive complexity caused by an excess of specificity. Semco was virtually buried under a vast quantity of policy, carefully stored in binders that went largely unread and unheeded but nevertheless impeded progress. The main use of these policy binders was to slow things down, reduce flexibility, and prevent decisive movement. What did Semler do? He threw them out. All of them—rules about office size, rules about work hours, rules about salary determination. Everything went out the window and was replaced with . . . almost nothing. No policy *is* the policy. Semler describes his company this way:

Semco has no official structure. It has no organizational chart. There's no business plan or company strategy, no two-year or five-year plan, no goal or mission statement, no long term budget. The company often does not have a fixed CEO. There are no vice presidents or chief officers for information technology or operations. There are no standards or practices. There's no human resources department.

And that's just a short list of what Semco doesn't have. What *do* they have? Passionate customers, engaged workers, and a sustainable cash flow. I'll take those over the other stuff any day.

It's important we don't take the wrong lesson here. Semler's strategy is not a form of organizational nihilism or anarchy, and the rest of us won't get the same results by simply torching the policy binders, then trying to carry on with business as usual. Like Semler, we have to replace the complexifying formalities with a simplifier that is even more powerful—trust.

Whether the complexity in our organization is rooted in size, structure, specificity, or something else, how we choose to relate to our partners will in large part determine our experience and our outcomes. Building relationships based on trust is infinitely simpler than formalized structures that dictate everything from work hours to dress codes and require people to repeatedly prove their competence and compliance with predefined norms. But trust is not just simpler—it is also better. This is not a moral opinion or a matter of taste. It's Science.

Research by Paul Zak, a PhD neuroeconomist at Claremont Graduate University, revealed that when people feel trusted, they respond by behaving in a more trustworthy manner because of increased levels of oxytocin in their brains. Oxytocin is the so-called cuddle hormone, which plays a critical role in a wide range of social interactions, including trustworthy behavior.

Zak's research suggests that if we want the people around us to be trustworthy, we can start by making them feel trusted. Rather than demanding that the people around us "earn" our trust, we would do well to give our trust from the outset and let neurochemistry do its work.

A 2011 paper in the *Journal of Trust Research* by R. C. Mayer et al. sheds even more light on the importance and value of trust. The paper describes a five-month longitudinal field study that showed that "the development of trust is a reciprocal phenomenon." This means one's trust in a given person is influenced by the level of trust one perceives from that person. In other words, if you trust me, I'm more likely to trust you. And lest there be any doubt, a 2011 paper in the *Journal of Public Administration Research and Theory* by Yoon Cho and Evan Ringquist showed that "trustworthy managers preside over more productive organizations and are better able to maintain and even increase organizational outcomes in agencies challenged by low level of performance."

But wait, there's more! Trust not only *increases* trustworthiness, it also *indicates* trustworthiness. Way back in 1980, Julian Rotter published a paper in *American Psychologist* that observed, "People who trust more are less likely to lie and are possibly less likely to cheat or steal. They are more likely to give others a second chance and to respect the rights of others. The high truster is less likely to be unhappy, conflicted, or maladjusted, and is liked more and sought out as a friend." Sounds good to me.

Rotter's research suggests that when someone trusts you, they are signaling something about their own character. If someone trusts you, it is a safe bet that you can trust them back. And yes, the inverse is also true—untrusting people tend to be less trustworthy. But Zak's research suggests that the virtuous cycle of trust is stronger because the brain's physical response to being trusted directly affects our behavior.

Sadly, despite the scientific basis and solid track record of trust-based teaming arrangements, some people find trust exceedingly difficult. They would rather trust a dusty binder and a five-year plan than the character or competence of the person sitting next to them. These nontrusters sling around phrases like "trust but verify," which usually means "I don't trust you until I verify everything." This is tragic and unnecessary.

According to Zak's research, this distrusting approach misses the point and has the opposite of the intended effect. Distrust creates a self-fulfilling prophecy in much the same way trust does. For men in particular, feeling distrusted causes a release of dihydrotestosterone, which triggers the desire for aggressive physical confrontation. This aggressive urge might explain the aforementioned predilection for drama that some people exhibit. Instead of physical violence, these dramatists express aggression by raising the stakes to extremes. Trusting them might therefore help to dial back the drama.

Zak explains that women are "cooler responders," adding, "we do not yet fully know the physiological underpinnings for this difference." Women may not respond to distrust by displaying aggression the same way men do, but the underlying dynamic is fairly universal. Regardless of gender, trust works. Distrust does not.

How do we make the transition to trust? At the risk of grossly oversimplifying things, we start by just doing it, perhaps starting with small things and building from there. To be sure, a genuine atmosphere of trust requires more substantial effort than just saying "I trust you," but that is a start.

Building a deeper sense of trust requires a certain amount of risk and exposure, a willingness to place one's well-being in someone else's hands, and a willingness to listen. Easier said than done, I know. But trust me—in the long run, trust works. See what I did there?

It is probably worth noting that there is a difference between being trusting and being gullible. Let's not mistake one for the other, refusing to trust out of a fear of being gullible or embracing gullibility in the name of trust. Instead, we can decide to trust wisely, not naively. For specific resources on how to foster trust, you may want to check out Henry Cloud's website at www. cloudtownsend.com/trust/ (full disclosure—Cloud and I have the same publisher).

Precious few of us have the authority to directly change the size, structure, and specificity of the organizations we're a part of. Even the CEO and chairman of the board can't easily transform a big company into a small one, a complex structure into a straightforward one, or a formal approach into a casual one. As for rewiring someone's personality to transform them from a drama-loving grenade thrower to a serene model of confidence and collegiality, well, that's not going to happen anytime soon, either. But we all have the capacity to be trusting and trustworthy. That just might be the spark that ignites the broader changes our organizations need. That just might be the first and most important step toward radical simplifications that make things better.

CHAPTER 6

The Doldrums, They Happen

Welcome to page 139. We are now roughly two-thirds of the way through the book and I would just like to say congratulations on making it this far. I'm told books like this one are purchased more often than they are read, let alone finished, and as much as I am thankful for the purchase, I sincerely appreciate the reading just as much.

The reason people often don't get this far in a book has little to do with the goldfish-level attention span of readers or the prodigious heft of the books themselves. The sad truth is, page 139 indicates the part of the book where many authors begin to phone it in (so I'm told). Why does this happen? Well, at this point in almost any book, the main idea has been presented and the critical evidence provided, but it's too soon for the grand finale. What remains is to bring the manuscript up to whatever scientifically calculated, totally not-arbitrary length is necessary to ensure bestseller status, usually via a series of further examples that basically restate earlier concepts or through fluffy little side notes that strive to at least provide entertainment if not insight.

It is generally felt that this part of the book is too late to introduce new material or concepts, because if said material were really important, surely the author would have brought it up already, right? Since new material is unwelcome at this point, all that remains is to restate earlier comments or introduce fluffy little side notes that strive to provide entertainment if not insight. Anyone else having a déjà vu moment here?

Lingering in this creative doldrums results in unnecessary additions that only make the book longer, not better. The result is a boring and long late-middle section that turns people off and causes them to either skip ahead to the conclusion or to abandon the book altogether.

Fear not, dear reader. I have no intention of complying with that pattern. I do, however, want to bring it to your attention, not for the sake of poking fun at my fellow authors or myself, but because this situation is not limited to literature. All sorts of design efforts can fall into the bland morass that so often typifies the late-middle portion of a book.

A project may begin with great energy, enthusiasm, and creativity, but at some point even the strongest among us find our momentum waning. We get tired, which makes us feel less focused and more distracted. This weariness nudges us toward familiar and comfortable behaviors rather than challenging ones, so we fiddle around the edges and continue numbly adding things and making small tweaks to the design instead of engaging with the more taxing effort of evaluating whether our additions and tweaks make things better.

Just when it is most necessary, the harder work of trimming and carving becomes, well, harder. The result—an unfortunate slide up and to the left, toward more complexity and less goodness. As we've already seen, that's the Complication Slope and it's bad news (there I go again, repeating myself).

When that happens, what to do, what to do? Coffee or naps will only get us so far, because the issue is not a lack of awakeness. This fatigue is more mental and emotional than physical, and even when our bodies feel well rested our minds can feel tapped out. The secret is to keep an eye out for this slowdown, to watch for it and prepare for it rather than have it sneak up on us.

This loss of creative vigor may be gradual or it may be sudden, and the timing may not be convenient, but the fact that it happens should not be a big surprise. If we are able to maintain an awareness of this predictable, virtually inevitable phase, we are less likely to get stuck in a counterproductively complexifying mode without realizing it. Fortunately, there are several ways to get out of the doldrums, and they all start by noticing that we're in them.

The first sign to look for is a loss of enthusiasm and a feeling of staleness. Our hands feel slow, our eyelids get heavy, and work that used to be exciting is now dull. That means we're in the center of the map and starting to creep toward the upper left area.

When we're in the doldrums, we also tend to shift the type of work we're doing—instead of adding new features and components because they make the design better, we're now making minor, low-value changes, changing happy to glad or painting the dark blue areas purple and the purple areas dark blue. If we can wake up our minds long enough to evaluate the impact of our actions, we just might notice that the impact of each change is negligible and the impact of all the changes together is negative. Yup, that's the doldrums.

Once we recognize what's happening, specific exit strategies will depend on how we got into the situation in the first place. If the project has been all-consuming for an extended period of time, the best move is to step away for a while and take that pause I mentioned earlier. No need to rehash that strategy entirely, but there is one particular type of pause that's worth mentioning. I call it the Shift Pause, because it involves shifting our attention and effort from one part of the project to another rather than stopping the work entirely.

This strategy works because creative projects are seldom monolithic and homogenized. Even the smallest project contains a variety of aspects and parts, and if we get bogged down on one part, maybe we just need to tackle a different part. For example, I wrote the first draft of this chapter when I realized I was fruitlessly spinning my wheels on an earlier chapter.

To be entirely honest, I couldn't stand to even look at that chapter anymore, and I wasn't doing anyone any favors by continuing to make imperceptible edits or adding new paragraphs that were little more than restatements of old paragraphs. Fortunately, writing about the doldrums in this chapter was a refreshing, cathartic diversion that simultaneously contributed to the overall goal of finishing the book and prevented me from making that other chapter unnecessarily worse. Shifting to work on this chapter was a pause from a part, not a pause from the work as a whole.

So pauses can come in many forms, long or short, partial or complete, active or restful. But sometimes a literal pause is exactly what we *don't* need.

If we're working on a stalled side project that feels stuck because of exhaustingly constant interruptions from our day job, then maybe the answer is to put *more* attention on the project, not less. Maybe we need the opposite of a pause and should instead accelerate things to the front burner for a while, to mix my metaphors. Personally, I always have a handful of side projects going at any given time, and when I find myself tiring of them and struggling to make progress, that's exactly what I do. One project in particular illustrates how this strategy works.

I was a junior engineer and it was the first time I ever initiated my own independent research project. It wasn't something I planned to do or something anyone asked me to do, but after reading books like Stephen Covey's *The Seven Habits of Highly Effective People* and Tom Peters's *In Search of Excellence*, I started to get ideas of my own. Specifically, a crazy little idea about work, goal setting, and risk taking took hold of my brain and wouldn't let go. I eventually gave it a name (*The Radical Elements of Radical Success*), but for the longest time it was just a short list on a torn scrap of paper I carried in my pocket.

For months, I felt as if I was on the verge of creating something big but I struggled to carve out sufficient time to put concentrated attention on it. I suspected my scribbled list was an iceberg tip of an idea, and I desperately wanted to dive below the waterline and see how big this thing really was. However, my day job was demanding, as day jobs often are, and I was also taking night classes for a master's degree. Plus, I was the proud dad of a one-year-old girl. Busy times, for sure.

So, I neglected the unofficial project because everything else was a higher priority. What else could I do, right? It nevertheless continued to linger at the back of my mind and the bottom of each day's to-do list, persistently tugging on my imagination. I occasionally managed to carve out brief interludes to dabble with it, but they always ended too soon and left me feeling torn, incomplete, and frustrated. Frankly, it was exhausting.

So I approached my boss, Joe, with a proposition: let me leave the office at noon on Friday for four weeks in a row, to go to a local library where I could do research and put focused attention on developing the idea. I promised the time away would not interfere with my formal duties, and I offered to report back on whatever progress I made. I was only asking for the equivalent of two days over the course of a month, but it felt absolutely extravagant.

He immediately and enthusiastically approved my request, solidifying his undisputed position of Best Boss Ever. During the next few weeks, I found the breakthrough I was looking for, in more ways than one. The project almost immediately fell into place and produced a methodology I used extensively over the next eight years. I developed a short training seminar to share the "Radical Elements" concept with my colleagues and delivered it several dozen times, for thousands of people, including college students, senior military leaders, and top officials from the intelligence community. Speaking to so many audiences not only allowed me to test and improve the idea; it also honed my skills as a speaker and expanded my network of collaborators, benefits I never envisioned during those hours in the library.

I have since moved on from that particular seminar, but it was a central aspect of my professional life for nearly a decade, thanks in large part to Joe agreeing to let me dedicate four Friday afternoons to it. The precious gift of uninterrupted hours was exactly what I needed.

The benefits of that focused time away extended into other areas of my life and work. Because I was able to commit a substantial, uninterrupted block of time to my independent research, I had more energy and creativity for my primary responsibilities. I was able to think more clearly all week long. Yes, I was doing 40 hours of work in 36 hours, but they were 36 focused hours and the results were better than ever. I even felt less distracted at home. Across the board, my life got simpler and better. Everybody wins.

Years later I asked Joe why he agreed to my request. He explained his philosophy that it is easier to direct energy than to create it. He knew that by saying yes to my crazy little project, I would go out of my way to do my main job well. Saying yes was a vote of confidence in my abilities and he knew my brief absences meant he would get more out of me each week, not less. He was also intrigued to see what my time in the library would produce and went on to be a vocal champion of the methodology I developed.

Not every boss is as enlightened and supportive as Joe, and not every job can accommodate that type of absence. We can't always carve out time on a Friday afternoon, much less four Fridays, and so that specific approach isn't always the answer, but it does serve as an example of a general strategy I like to call Make the Time.

Depending on our situation, we might have to get more creative in our effort to find those precious hours. Writing this book, for example, mostly happened at 5 a.m. I didn't need special permission to carve out these early hours and dedicate them to putting words on paper. All I needed was a loud alarm clock and a million pots of strong coffee.

The nice thing about this approach is that it didn't take time away from anything other than sleep. I seem to need less sleep than most people to begin with, and I cleverly managed to make up for the missing rest by promptly nodding off in the middle of whatever sitcom or police drama I'd be watching at 10 p.m. A word of caution if you're thinking about adopting this strategy, however. Researchers tell us that inadequate sleep has many undesirable side effects, including heart disease, weight gain, and your spouse throwing a pillow at you when you start snoring in the middle of *Castle* again. Fortunately I only encountered one of those side effects, but the point is we should not skip too much sleep. Also, be sure to eat your vegetables and get some exercise.

But spending too much or too little time are not the only paths into the doldrums. Sometimes the issue has nothing to do with hours and everything to do with partners.

When a solo project hits this phase, it might be time to bring in a buddy or two to lend a hand. Maybe we need help from someone with a whole new skill set, or maybe we need someone with a deeper understanding of the skills we are using. For example, a computer programmer might reach out to a more experienced mentor who can help identify a more productive approach to writing code. Or she might team up with someone who knows nothing about programming and instead talk about what the program should do rather than how it should do it. Either way, the benefit comes from having a fresh set of eyes.

Alternatively, if we've got a large team already and the project is stuck in a complexifying morass where progress dwindles to a trickle, it's probably time to temporarily break away for some solitary cogitation.

Nobel Prize–winning physicist Richard Feynman had a habit of doing precisely that. He worked on some of the twentieth century's most prominent physics programs, from the Manhattan Project to the *Challenger* disaster investigation. Each time, he tackled very difficult problems with a large team of collaborators. However, his biography by Jim Ottaviani and Leland Myrick shows him frequently spending time alone, usually engaging in activities that were entirely unrelated to the task at hand—learning to be a safecracker, playing the drums, or pursuing one of his other quirky hobbies.

This behavior was not the result of a merely casual interest in recreation, although by all accounts Feynman took great pleasure in finding things out. Sure, he was having fun, but his habit of going away from the group was also a deliberate problem-solving strategy. In a letter to fellow Nobel Prize winner Enrico Fermi, Feynman explained, "I get lots of ideas at the beach." He was by no means a solitary scientist, but he regularly made time to get away, to be alone, and to think before returning to the larger group. These trips to the beach tended to correspond with situations where he was stuck on a particularly thorny challenge.

The common element in all these strategies is the change in perspective. Whether we're going from full-time to stop, part-time to full, solo to team, or team to solo, having a new angle on the work is key to breaking out of the inevitable doldrums. We might even want to plan these adjustments in advance, setting up relationships or scheduling time periods where we can shift our attention appropriately.

No doubt there are other reasons we find ourselves in the doldrums, and thus other strategies for getting out of them. Maybe the problem is that we're working on the wrong project altogether, in which case the pause should be permanent after all. Maybe we need to get away from the screen and spend some time with a pen and paper. Maybe we need to get out of the office and spend some time in the woods.

Regardless of the specific reason or the specific solution, one thing to keep in mind is that the doldrums happen and that's something we should prepare for. When they do, we need to notice they are happening, then find a way to get some new perspectives and avoid allowing our momentum to carry us in the direction of increased complexity and decreased goodness.

CHAPTER 7

Travelogues and Archetypes

ike most of my favorite books, my copy of Donald Schön's *The Reflective Practitioner* is well worn and rigorously underlined. I have read Schön's masterwork on "reflection-in-action" many times over the years, each time leaving myself a trail of inky bread crumbs as I highlight certain sections or scratch a brief (usually cryptic) note to my future self in the margins.

I make it a point to flip through its pages on a regular basis, rereading the underscored portions and finding new passages to underline. Sometimes I go looking for a specific phrase or section, while other times I simply read to rediscover and remember. Without fail, each time I open this book I come away with a new thought to reflect on and apply.

In a recent excursion into *The Reflective Practitioner*, I was struck by Schön's observation that "professional practice has at least as much to do with finding the problem as with solving the problem found. . . ." He goes on to write that for most professionals, "well-formed instrumental problems are not given but must be constructed from messy problematic situations."

Schön is not the first or only writer to emphasize the importance (or difficulty!) of problem discovery, but his description resonates strongly with me. As I wrote my own book, it was a helpful and timely reminder that all the attention we put on problem-solving techniques amounts to nothing if we end up solving the wrong problem. Finding the right problem is therefore a primary objective in almost any professional endeavor and meaningful progress starts by understanding what is meant by the word *good*.

In fact, bad definitions of *good* are a central problem from which all sorts of other problems emerge. This is why the Simplicity Cycle emphasizes "increasing goodness" as the objective of our efforts and encourages us to spend some time considering what goodness really means, rather than contentedly pursuing simplicity in all situations. If we merely make things simpler, we risk solving the wrong problem.

How, then, can we make sure we have found the right problem and are pursuing the right kind of goodness, particularly when we face a "messy problematic situation"? The discipline of reflective practice is certainly helpful in that situation. So is a map. Put the two together into a "reflective mapping" activity and we've really got something.

The Simplicity Cycle provides a framework to do precisely that. It maps out the relationship between a variety of behaviors and outcomes, and invites us to consider the many paths we might follow. In doing so, it makes it easier for us to correctly identify complexity-related problems . . . and then to reflect on them, avoid them, address them, resolve them.

To help foster such reflection, the following pages present a collection of archetypical journeys through the Simplicity Cycle, each accompanied by a short commentary. These mini-travelogues are snapshots from the road, a sharing of thoughts, hints, and tips from one traveler to another. This is not a definitive collection by any means, but instead is a starting point for additional conversations and an example of what reflective mapping and problem finding might look like.

Using the Simplicity Cycle as a framework for recording and comprehending our experiences can lead to helpful insights. Readers may want to create sketches of their own in order to better understand the road behind them and prepare for the road ahead. No fancy equipment is needed, just a pen and paper or perhaps a whiteboard. We begin by laying down a blank Complexity-Goodness field, then think about how our project began and how it proceeded. What steps did we take to make things better? What steps did we take that made things worse? What paths did we follow, and where did these paths lead us?

This can be a communal activity, with a large group or a small one. It can also be an individual exercise. These are not mutually exclusive options, and a team may find it useful to transition between solitary reflection and group reflection. In terms of formality, this can be an idle doodle or a disciplined practice.

As we sketch and reflect, we do well to keep in mind that any given journey is part of a larger cycle. Although we may happily arrive at a finish line and produce something that is very good, the future always holds new horizons and endless opportunities for improvement. These little maps may tell the stories of long trips and triumphant arrivals, but each one is only a chapter in a story with no end.

STUCK ON THE PORCH

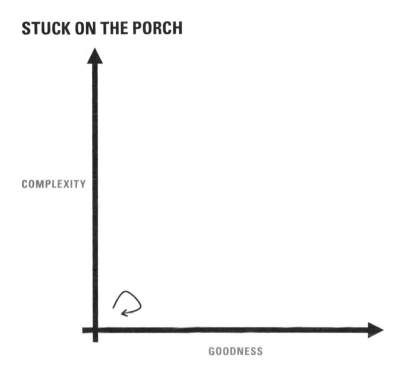

FIGURE 24: STUCK ON THE PORCH

The journey of a thousand miles may begin with a single step, but taking that first step can be harder than it sounds. Whether it's a series of false starts or tentative attempts that are immediately scrapped, we sometimes get stuck on the front porch, unable to head out into the world. Our design is neither complex nor good, because it hasn't really begun.

In this situation, ideas pop up and are instantly dismissed as unworthy of being included, resulting in a stubbornly persistent blankness on the page before us and a growing pile of balls of crumpled-up paper in the bin beside us. Even worse, ideas refuse to come. The page stays blank and the bin stays empty.

This is different than the doldrums, which occur in the middle of our efforts where we feel bogged down by all the moving and working and trying lots of things that simply don't work. Getting stuck on the porch happens on day one and prevents us from getting to day two.

True story: this particular section of prose spent an ironically long time stuck on the porch. If you could see my drafting notebook for the preceding pages, you would see a small set of miserably bad ideas that went nowhere and were crossed out as soon as they appeared. What my notebook does not show is the long stretches of time between each cross-out, where I stared blankly at the page and had no words to add, nor does it show the even worse ideas that briefly came to mind but never quite made it to my pen.

Why does this happen? Because we get ahead of ourselves and refuse to tolerate the messy side of the process, where half-baked ideas must come (and go) as we experiment and sift through the chaff. When we expect every word we write to be correct, every line we draw to be smooth, and every piece we add to elegantly complement the whole, there is no space for being wrong. We apply a standard of excellence that has no place at this stage of the journey, a standard that paradoxically prevents the very quality it is supposed to enable.

The solution is to fire your inner editor and give yourself permission to do bad work. Anne Lamott colorfully encourages writers to produce Shitty First Drafts, a practice that is relevant far beyond the literary arena. In a similar vein, G. K. Chesterton argued that a thing worth doing is worth doing badly. And Winston Churchill wrote about the wisdom of grabbing a large paintbrush and boldly, fearlessly splashing color onto a canvas, without concern for whether or not the first pass produces the desired effect.

Easy for them to say, right? Not at all. This was surely a hard-won lesson for each of them. But easy or not, they are all correct, and dismissing their advice because "it's more complicated than that" is a cop-out and a lame excuse. Our difficulties are no greater than theirs and should not be allowed to prevent us from getting started.

Yes, the first step is scary. Yes, the first step is hard. Yes, the first step is often wrong. That is all irrelevant. What matters is that the first step is *essential*, and if we don't take it, then we can't go anywhere. The first step may not take us in the right direction, but that's okay. At this initial phase of the journey, direction matters far less than movement. Course corrections can come later, once we have a course to correct.

LUCKY

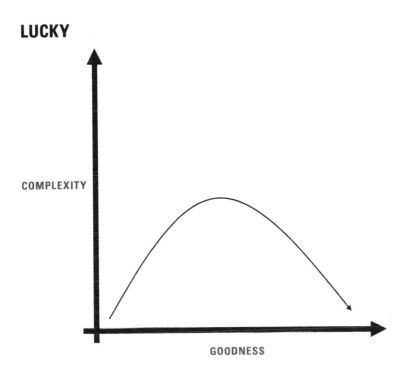

FIGURE 25: LUCKY

Sometimes, we jump right off the porch and hit the ground running. Sometimes, the stars align and the muse is generous. Sometimes, we immediately experience what Mihály Csíkszentmihályi refers to as a state of *flow*, completely absorbed in a challenging activity that activates our creative powers to the fullest.

We progress smoothly and directly from one slope and phase to the next, with no hiccups or detours along the way.

Sometimes, we get lucky.

The inventor of the Slinky was a lucky U.S. Navy engineer named Richard James. In 1943, he was experimenting with prototypes of springs designed to hold sensitive instruments securely in place at sea. According to legend, James accidentally knocked one over and watched it tumble in the now-familiar Slinky motion, inspiring him and his wife, Betty, to create a beloved childhood toy that went on to sell more than 300 million pieces.

While the historical details of the Jameses' story are no doubt more complex than the legend, the overall trajectory of his invention looks much like Figure 26. The resemblance of the final design to the original shipboard version might conceal the effort involved from a superficial investigation (for example, the couple spent two years experimenting to find the best steel gauge and coil size), but to experienced eyes, the Slinky's final design reveals the results of a determined effort to reduce complexity.

COURSE CORRECTION

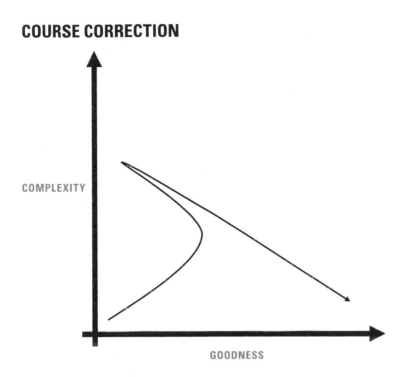

FIGURE 26: COURSE CORRECTION

But sometimes we are less lucky. Sometimes we end up in a place we never intended.

When the complexity of a design overwhelms its goodness, the solution is to make a course correction. Streamline, simplify, integrate, and remove the unnecessary bits.

Frankly, it almost doesn't matter which bits we remove first. Start anywhere. Remove anything. When a design is in that upper left quadrant, almost any trimming will make the thing better. Keep trimming long enough and we might end up in a very good place indeed.

In an 1890 letter to the Royal Society, Australian aviation pioneer Lawrence Hargrave described his approach to building experimental aircraft models. While initial efforts to build heavier-than-air flying machines in the pre–Wright brothers era were massively complex, Hargrave explained that he had "swept away such a mass of tackle from the machine that its construction becomes a ridiculously simple matter."

Octave Chanute praised this approach in his 1895 book, *Progress in Flying Machines*, pointing out that Hargrave "marked a very considerable advance in design by a great simplification of the propelling arrangement" and writing, "It will be noted that the engine is a marvel of simplicity and lightness." This was a radical departure from the complex designs of the era, an important and influential course correction.

After his breakthrough in simplified engine design, Hargrave began investigating different geometries and shapes for producing lift. In order to keep his efforts focused, he used kites with no engines at all instead of actual airplanes, and eventually produced box kites capable of lifting people into the air. Far from a step backward, these simple kites greatly advanced our understanding of how to overcome gravity and helped define the shape of wings for generations to come.

Chanute summarized Hargrave's contribution to the development of aviation with these words: "Thus with small, light, simple, and inexpensive models many experiments were made, and great advance realized. . . ." Hargrave showed the way to an important course correction and helped build the foundation from which Orville and Wilbur Wright would rise.

OVERCORRECTION

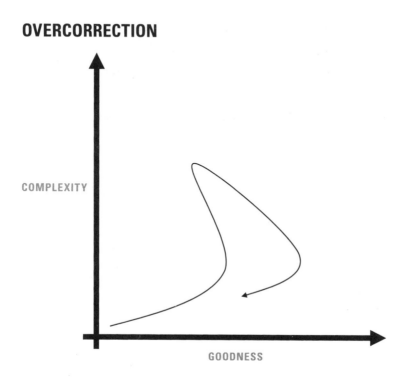

FIGURE 27: OVERCORRECTION

Did I say "almost any trimming will make the thing better"? Yes. But if we try hard enough, we can certainly do it dumbly and end up heading in the wrong direction after all.

Suppose a well-meaning legal reformer wanted to reduce drug-related offenses but felt that the current system of laws, trials, appeals, and loopholes was too complicated, slow, and expensive. If our good-hearted innovator decides to pursue simplicity at all costs, he or she might be tempted to propose a radically simple solution: mandatory and immediate death sentences for anyone caught selling, holding, or using illegal drugs. This would obviously bring the recidivism rate to zero. There would be no more messy appeals or retrials. Judges and juries would scarcely have to think at all—if the accused is guilty of being in the vicinity of drugs, off with their heads! What could be simpler?

The only problem with such a scheme is that it is horrifyingly terrible and is the exact opposite of justice. While the approach is indeed simple, the failure to distinguish between virtuous and unvirtuous simplicity helps explain why the concept is so very, very wrong.

Remember, the goal is to make the design *better*, not just *simpler*. One more time: simplicity isn't the point. If we overvalue simplicity, we might get distracted and end up overcorrecting, cutting to the bone and reducing complexity long after such reductions cease to be wise or productive. Course corrections are important, but we must be careful not to lose sight of the objective.

RETURN TO BASECAMP

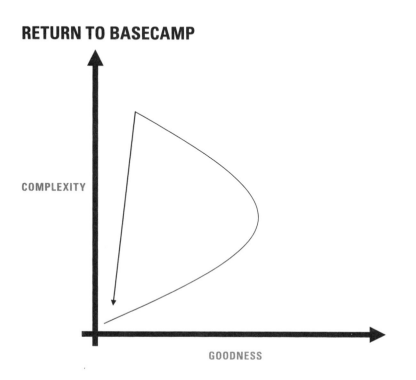

FIGURE 28: RETURN TO BASECAMP

Complexity can be a mental trap, ensnaring our imagination and restricting our ability to see simpler possibilities. Upon creating a monstrously complicated beast, the prospect of simplifying it may be wholly overwhelming.

Where do we start? Which thread do we pull first?

The beast snaps at our fingers each time we approach, resisting even a modest attempt to trim its beard. Surely it will not submit to a more thorough examination. Surely it will not accept restraint.

The solution may require bolder action than a straightforward course correction. It may require a fresh start.

Scrap it. Go back to the beginning. Start over.

Note that in Figure 28, the significant decrease in complexity diminishes the design's goodness a bit. It is only a slight decrease, but we have indeed taken a step backward along the Goodness axis and our design is now (slightly) worse than it was before. This slide to the left is okay because our overall posture and position are so much improved. The design is unburdened and so are we, which makes it easier for us to set off in new directions that were previously hidden by excessive complexity. A small step backward prepares us for a big leap forward.

This can be a difficult strategy to adopt, in part because large, complex designs are often the product of many contributors, each of whom feels protective of their particular addition, whatever they might think of the design's overall effectiveness. These contributors may seek to retain or reintroduce their favorite piece of the design, and if every piece is someone's favorite, we'll end up back in the upper left corner again.

Even on a solo project, restarting can be difficult because of our tendency to reapply our first approach on each subsequent attempt, resulting in versions 2.0 and 3.0 that look suspiciously like version 1.0.

Fortunately, with a little discipline and effort, starting over is not as difficult as it looks. In fact, it can even be quite fun. And yes, it's okay for the new design to retain some elements of the old. The key is to make sure we keep only the most important and impactful pieces.

When Jason Fried and his colleagues at software company 37signals changed the company name to Basecamp in 2014, they weren't exactly scrapping *everything* and starting over, but it was pretty close. The new name reflected their decision to reduce the company's offerings to their single most popular product (a program management app called Basecamp) and to step away from other software tools they had developed over the past fifteen years, with names like Campfire and Highrise.

The Basecamp website explains that the company is "doubling down on simplicity," a logical extension of their longstanding design philosophy. As 37signals, they had always made a point to "start with no" when considering new features, resisting unnecessary additions and complications to their software. Now they were applying that No to the company itself, returning to an earlier degree of simplicity, where they had fewer products but provided better service.

Did this move really bring them all the way back to the lower left corner of the chart? Maybe, but if so it was only for a brief stay while the company and its customers recalibrated their definition of goodness. For that matter, we might ask whether the company was ever really in that upper left corner. If we compare them to their competitors, the answer is almost certainly no. But if we compare them with their own ideals, the answer just might be yes, and that is the comparison that matters most.

COMPLACENCY

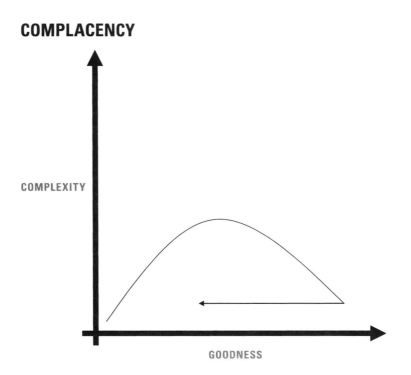

FIGURE 29: COMPLACENCY

The great English journalist G. K. Chesterton wrote:

> If you leave a thing alone you leave it to a torrent of
> change. If you leave a white post alone it will soon be

a black post. If you particularly want it to be white you must be always painting it again. . . . Briefly, if you want the old white post you must have a new white post.

Left alone, a simple design does not become complicated. However, a good design exposed to a torrent of change eventually becomes a bad design. The transition from good to bad does not mean the design itself is any different than it was yesterday. In fact, it hasn't changed at all . . . and that is the problem. While the design stayed static, the world changed around it and what once looked crisp and white is now rather dingy.

If we want to maintain the old goodness, we must continually pursue new goodness. We cannot rest comfortably on our laurels for long, as if the law of entropy did not apply to us. If we want the post to stay white, we have to continually paint it white. If we want our design to stay good, we have to continually work to make it good.

Clayton Christensen's concept of disruptive innovation comes into play here. Successful, well-established firms with a track record of delivering good products are particularly vulnerable to the threat of disruption in the form of "technologically straightforward" products that are "often simpler than prior approaches." These innovations introduce change to the market environment that status quo defenders are unable or unwilling to address. The result is a marked decrease in a static design's goodness relative to the dynamic world around us.

How can we avoid or resolve the dilemma? Christensen has published several books on the topic and I won't attempt to summarize his solution in a paragraph or two. Instead, I will simply suggest that the first step toward a solution is to strenuously avoid complacency. The second step is to read his books.

LOOPY

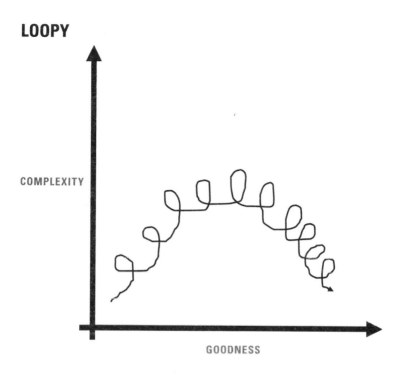

FIGURE 30: LOOPY

Perhaps a Zen master can proceed through the cycle in a straight line, undistracted and laser-focused on the goal. But the mortals among us are more likely to do something that looks like this.

The loopy path may not be the most efficient way through the cycle . . . but on second thought, maybe it is. Each small loop is a sign of exploration and learning, so this pattern represents a path of change, discovery, and correction. The stories of such loops are seldom told or recorded, because each little circle is a modest excursion that leads to another and is thus quickly forgotten.

By keeping the loops small we constrain the costs involved (financial and mental), but even a series of larger loops may depict genuine progress.

Perhaps this is what a Zen master would do after all.

WANDERING

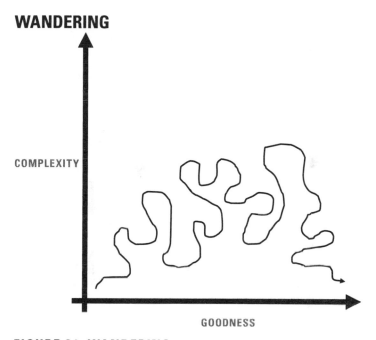

FIGURE 31: WANDERING

J. R. R. Tolkien wrote, "Not all those who wander are lost." As we explore new design territory we must sometimes meander in order to find our way.

The experience of design, the journey of design, is seldom linear and predictable. If our path looks something like Figure 31, it might just mean we are staking out undiscovered territory.

That might be the best kind of journey.

GETTING LOST

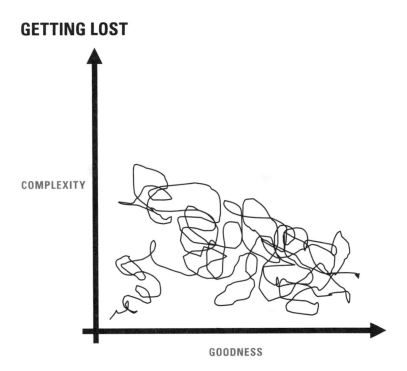

FIGURE 32: GETTING LOST

Tolkien was correct that not all those who wander are lost, but some wander because they truly have no idea who they are, where they are, or where they are going. To be honest, sometimes "they" is "me."

What can we do when the North Star is hidden, the trail

markers are absent, and every street sign is in a language we don't comprehend?

There are several strategies to resolve this situation. The Boy Scouts tell us that if we get lost in the woods we should stay in one place and wait for rescuers to find us. This is sometimes an option.

Winnie the Pooh offers different advice—if you keep walking past the same sandpit while trying to find your way home, perhaps the solution is to look for the sandpit instead and thus find your way home.

I can't promise that approach will work for you, but it worked for the bear and his friends.

OPTIMAL RANGE

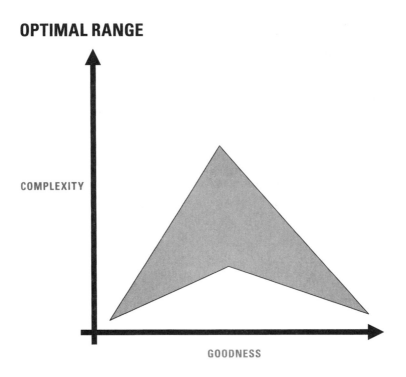

FIGURE 33: OPTIMAL RANGE

Whether you are looping, wandering, or trying to get home again, one thing to keep in mind is that the thin, straight lines used throughout this book represent a notional path, a stylized tracing of the design process.

Deviations from the straight and narrow are to be expected

and might even be desirable. Those little excursions could lead to breakthrough discoveries.

So perhaps we might think in terms of staying within an optimal range rather than staying on a narrow trail. This hopefully reduces any pressure we might feel to accurately calculate our exact position on the chart. As long as we keep to the shaded region and our overall trajectory is pointed in the direction of increased goodness, we're probably doing just fine. The fact that this loosely resembles a *Star Trek* logo is just a happy coincidence. I promise.

PROFESSIONAL PROGRESSION

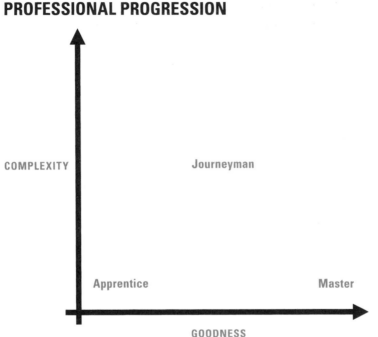

FIGURE 34: PROFESSIONAL PROGRESSION

Across a career or a lifetime, we will set up a series of residences in different regions, moving on to a new destination after learning the lessons of the old place. The transition from apprentice to journeyman to master describes one such progression.

When we begin a new craft, we are an apprentice, laboring in the region of simplicity. An apprentice approaches a trade with little knowledge and even less ability. The apprentice's task is to gather information, skills, and techniques, expanding their repertoire and capability. When the apprentice looks ahead, he or she sees a mountain of learning to be climbed.

A journeyman's situation is different. Having studied at the feet of a master, a journeyman now has a full toolbox. It may not contain every tool they will ever use, but the vast majority are indeed present by this time. For journeymen, the most productive learning and most important lessons involve deepening their understanding of when and how to apply their tools, rather than adding new ones. Wisdom also comes from learning when and how to *not* use a particular tool.

The master works in simplicity, years of experience allowing him to perform each task with an elegant economy of effort. A master's hands use tools with minimal strain, maximal precision, and grace. And of course a master must simplify the craft enough for an apprentice to understand. This completes the cycle.

IGNORANCE TO WISDOM

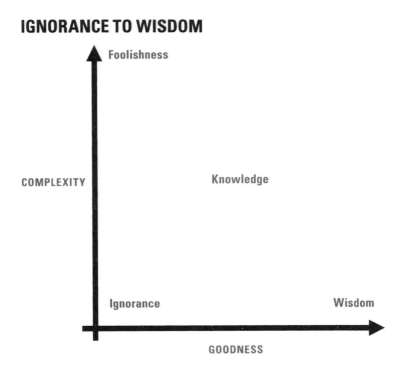

FIGURE 35: IGNORANCE TO WISDOM

Like an apprentice, we often begin in ignorance. Without experience or information, our minds are empty vessels, awaiting fullness.

And so we learn. We explore. We study. We experiment. We accumulate information and it becomes knowledge. This is good.

Until it isn't. Knowledge isolated from goodness loses its value, as we learn more and more but understand less and less. All the great spiritual teachers, from Solomon to Jesus to Buddha, attest to the way knowledge can descend into foolishness. The problem isn't with knowledge itself, but with the way we tend to idolize it, cherishing it above its value.

In contrast, wisdom involves seeing the patterns, seeing the connections, integrating information into meaning, and recognizing the limits of knowledge. Wisdom is about reduction, not accumulation. It is about discerning what is good and relevant and not being distracted by ephemera.

Wisdom is about simplicity. The master, too, may be an empty vessel.

REACTIONS TO COMPLEXITY

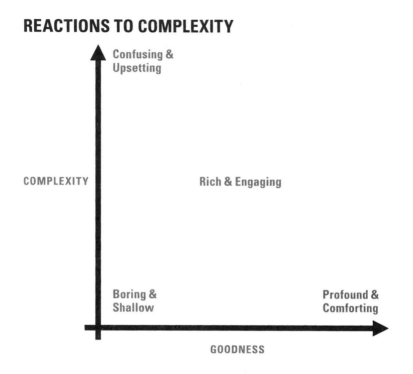

FIGURE 36: REACTIONS TO COMPLEXITY

For all the nice things we've said about simplicity, sometimes that's not what we really want.

In the middle of the chart we will find devices, objects, and experiences with multiple, interconnected layers. For example, a mystery novel with a large cast of characters and frequent plot twists is fun to read precisely because of its complexity. An orchestral performance uses layers and complex themes to engage our hearts and passions. A video game immerses us in a dazzling world, full of complex challenges and surprises.

Yes, we can appreciate simplicity in each of those contexts—a simple mystery, a simple tune, a simple game. But sometimes the most enjoyable instances would never answer to the description "simple." Sometimes the thing we're looking for belongs, properly, in the center of this chart, and the complexity contributes significantly to the listener's/player's/reader's engagement and enjoyment.

In creating such rich complexity, the trick is to ensure that each clue, each character, each note, and each instrument contributes to a coherent whole. For a complex creation, it is particularly important to remove the discordant, distractive elements, while retaining the interesting depth.

Of course, the *resolution* of the story, song, or game occurs in the lower right corner, where the pieces fit together in a neat conclusion. It is perfectly okay to let an audience linger in the middle of the chart for a while, but by the time the show is over, we should make sure they are brought safely out.

MODEL T AND VW BUS

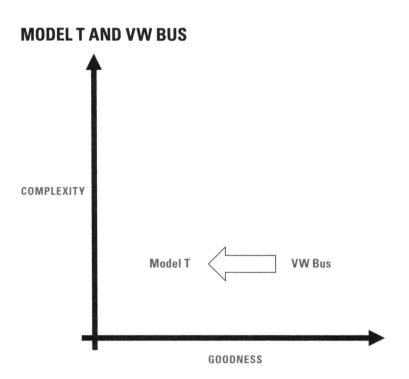

FIGURE 37: MODEL T AND VW BUS

The Model T's success was largely the result of its simplicity. Henry Ford's assembly lines reduced the complexity of manufacturing by breaking the experience down into small, manageable steps using interchangeable parts and common standards. The car itself was simple enough for an ordinary person to drive and maintain it.

In the early twentieth century, the Model T would have resided squarely in the lower right corner of our chart. But time presses onward and nudges even the best product to the left, in the direction of decreased goodness.

By the 1960s, the Volkswagen Bus enjoyed a similar status of residing in the lower right corner. It was a wildly popular vehicle, precisely because of its simplicity and reliability (the two are almost always directly related). The old Model T had not become more complex. It just wasn't as good as it used to be, because the market had changed. Tastes had changed. Technology had changed.

And no, the VW Bus didn't get to stay in the lower right corner forever. Nothing does.

A FINAL EXAMPLE

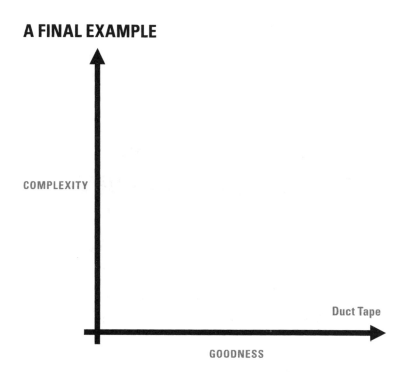

FIGURE 38: DUCT TAPE

'Nuff said.

FINAL THOUGHTS:
ON MAPS AND JOURNEYS

William Least Heat-Moon's road trip memoir *Blue High-ways* ends the way any epic journey should, with the hero's return. After driving his tricked-out van on a thirteen-thousand-mile loop around the continental United States, Least Heat-Moon fuels up one last time and begins the final leg of his journey, back to Missouri, where the adventure began.

Well, sort of.

After three months on the road in 1978, the man who returned was not the same as the man who departed. Maybe the difference is in the way he walks. Maybe in the way he stands. Maybe it's in his eyes. Or maybe it's everywhere and nowhere all at once. The change is subtle and impossible for an external observer to quantify, but it is surely profound and almost certainly permanent.

His hometown no doubt looked much the same to those who never left, but it, too, had changed. Least Heat-Moon's new eyes saw it differently than they had before, and in a very real sense, the home he came back to had been re-created as profoundly as Least Heat-Moon himself. For that matter, so had the entire universe. If you've ever been on an adventure, whether it's around the block or around the world, you probably know what I'm talking about.

A similar change happens on a journey of design. We set out into territory that may be familiar or may be alien, along well-trod paths or through virgin forests, along blue highways or red ones. But regardless of novelty or difficulty, whether the road is smooth and straight or rough and winding, we are remade by the experience. Transforming a blank sheet of paper, an empty screen, or an unshapen lump of clay into something new cannot help but transform us as well, and this transformation will affect all our future journeys.

With our new eyes we see the world around us in a different light. We are aware of previously hidden alternatives and open to new opportunities. We come away with a better understanding of why previous designers made certain decisions. We may even discover paths that take us to better places than our predecessors inhabited. Perhaps we learn that a comfortable, familiar path does not take us where we wanted to go after all, or that a foreboding doorway leads to a more beautiful place than we could have imagined. We may learn that the old map was wrong or incomplete. And if we are particularly bold and curious, we might even be able to fill in empty spaces on the map, as a boon to future travelers.

These discoveries shape our subsequent interactions and decisions, our additions and subtractions. They affect our definition of goodness and our appetite for complexity. This changes everything.

The secret ingredient that catalyzes all this change is the road dust on our own boots, the sign that we have actively engaged with the world rather than remained a bystander. When we replace the old dust of inaction with new dust from the trail, we gain a special kind of power and a special kind of authority. We speak with credibility when we say, "I was there . . ." even if we're only speaking to ourselves.

A good map is one of the most useful tools we can take on a journey. As with any tool, learning to use a map is a skill that requires guidance, practice, and effort. A tenderfoot's initial exposure to a map produces bewilderment rather than insight. The symbols and shapes mean nothing at first. But once an experienced guide explains how to read the key, how to orient the map, and how to interpret the various lines, squiggles, and shadings, meaning emerges. The beginner can now find his or her way, and may even visit places the guide never knew about.

A map does not only show us one place or equip us for a single journey, it shows us the whole world, a lifetime of journeys. Maps can thus be a source of inspiration, showing that places exist that we cannot see from where we currently stand. William Least Heat-Moon looked at his map every day and picked as his next destination a town with an interesting name, connected by the thin blue line of a small highway rather than the thick red line of the interstate. Although these towns were not directly visible from his starting point each morning, he could see their names on the map. He could see which route would take him there. And that was enough to make the entire trip possible.

But even the best map is incomplete. Recall from chapter 1 that the map is not the territory. It cannot tell us what the weather will be like, how thick the vegetation will be, nor the names of the people we might meet along the way. The closeness of lines on a topographical map suggests the ache we will feel in our legs as we scale a steep hill and the excitement of reaching the summit, but the only way to experience the view is to make the trip.

The Simplicity Cycle is thus a starting point, a potential source of inspiration for your design efforts. I hope it helps distinguish the smooth paths from the rocky dead ends. I hope it helps equip you for many fulfilling expeditions, as you explore and experience the cycle. Of course, reading this little book is no substitute for putting pencil to paper (or fingers to keyboard or chisel to stone) and actually designing something. That is where the real change begins.

So by all means, spend some time with the map. Study its contours and twists. Dream about voyages to strange new lands. Brace yourself for the rigors of the road. Then head out into the unknown and make something beautiful. When you come back home, there is a pretty good chance you'll find that you've changed, and so has the world.

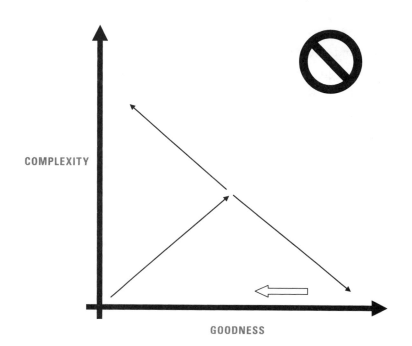

FIGURE 39: THE SIMPLICITY CYCLE

ACKNOWLEDGMENTS

I owe an enormous debt of gratitude to so many people who helped make this book a reality. At the risk of oversimplifying things, here is a brief story of how it all happened. My thanks go out to each person named below.

The Simplicity Cycle began as a sketch in a notebook, inspired by a conversation with Lieutenant Commander Jenny Soto from the U.S. Navy. I hashed out the idea further with a whiteboard and a small group of supremely talented Air Force innovators at the Rome Research Site in central New York, including Brian Yoshimoto, Gabe Mounce, Kurt Barsch, Ryan McKeel, and Kevin Bartlett, most of whom were captains at the time but went on to earn several further promotions.

The first published version appeared as an article in the November–December 2005 issue of *Defense AT&L* magazine, with editorial help from the inimitable Judith Greig. I then published a longer version at ChangeThis.com. At Dan Pink's suggestion, I started turning this funny little idea into a book. With Don Norman's enthusiastic encouragement I continued refining it, making it simpler and better. Don also introduced me to his wonderful agent, Sandy Dijkstra, and her colleagues at SDLA, including Elise Capron and Roz Foster. They took me on as a client and secured a book deal with HarperBusiness, where Hollis Heimbouch worked her brilliant editorial magic and helped me remake, reimagine, and reframe the manuscript into something far better than I ever dreamed it could be. Every writer should be so lucky to work with Hollis.

Along the way, I received endless encouragement and assistance from colleagues and students too numerous to count. My friends Will Goodman, Lieutenant Colonel Matt Keihl, and Lieutenant Colonel Chris Quaid all helped me more than they know. Professor Dave Barrett from Olin College was particularly supportive, posting the Cycle diagram on the wall in his lab and regularly inviting me to speak with his students. Lieutenant Colonel Heino Matzken from the German Army confirmed that I was using *verschlimmbesserung* correctly. Rolf Smith, aka Colonel Innovation, graciously allowed me to share it on a School for Innovators "Thinking Expedition" and with the Association of Managers of Innovation during an incredible weekend in Death

Valley, California. And even though our all-too-brief time working together ended a few years before I came up with this idea, Lieutenant Colonel Joe Wotton deserves much credit for helping me build the foundation as a writer, thinker, and engineer upon which this book is built.

Throughout this journey, my wife, Kim, has been a loving companion and a source of much strength, wisdom, and blessing. Our daughters, Bethany and Jenna, bring tremendous joy and fun into my life. As always, this book is for you three.

GLOSSARY

Complex: Consisting of interconnected parts. Objects can have high or low degrees of complexity.

Complexity Slope: A phase in which complexity and goodness are increasing proportionally, represented by a line running up and to the right. The primary activities along this slope are additive, as new features, parts, and functions are introduced to the design. These additions produce a positive effect on the design's quality.

Complicated: A state in which complexity is high and goodness is low, as represented by the upper left corner of the Simplicity Cycle diagram.

Complication Slope: A phase in which complexity and goodness are inversely proportional, where complexity increases and goodness decreases. The primary activities along this slope are additive, as new features, parts, and functions are introduced to the design. These additions produce a negative rather than positive effect. This slope is perpendicular to the Complexity Slope and runs up and toward the left.

Cycle: A repeated series of activities.

Doldrums: A phase characterized by fatigue, apathy, and confusion in which progress is slow or nonexistent. Usually occurs in the middle of a long project.

Efficient: Containing precisely the right quantity of parts.

Goodness: A general term that represents the measures of merit and desirable attributes of a particular object or design.

Make the Time: A strategy for breaking out of the doldrums, which involves strategically investing time to achieve top-priority objectives.

Pause: A temporary break in work, for the purpose of breaking up unproductive momentum and allowing a mental reset.

Premature Optimization: An ill-timed shift from the Complexity Slope to the Simplification Slope, instituted before assembling a sufficient mass of complexity. The resulting product provides minimal goodness.

Shift Pause: A pause strategy that involves shifting activity from one component or aspect of a design onto a different component or aspect, rather than a suspension of activity entirely.

Simple: A state in which complexity is low, found on the lower half of the Simplicity Cycle diagram.

Simplification Slope: A phase in which complexity and goodness are inversely proportional, where complexity decreases and goodness increases. The primary activities along this slope are reductive, as new features, parts, and functions are subtracted from the design. This slope is perpendicular to the Complexity Slope and runs down and toward the right.

Simplistic: A state in which complexity and goodness are both low, found in the lower left corner of the Simplicity Cycle diagram.

Special Piece: A component that, when added to a design, drastically improves the design's goodness and facilitates removal of previously added components.

Trimming: Removing a part (or series of parts) from a design, in order to identify which elements are essential and which are extraneous. This practice comes from the TRIZ methodology.

TRIZ: The "Theory of Inventive Problem Solving," a methodology invented by Russian inventor and scientist Genrich Altshuller.

Verschlimmbesserung: German word for an improvement that makes things worse.

SELECTED SOURCES

Chanute, Octave. *Progress in Flying Machines*. Long Beach, CA: Lorenz & Herweg, 1976.

Christensen, Clayton. *The Innovator's Dilemma: The Revolutionary Book That Will Change the Way You Do Business*. New York: HarperBusiness, 2011.

Dertouzos, Michael. *The Unfinished Revolution: Human-Centered Computers and What They Can Do for Us*. New York: HarperCollins, 2002.

Gladwell, Malcolm. *Outliers: The Story of Success*. New York: Little, Brown, 2008.

Lackoff, George, and Mark Johnson. *Metaphors We Live By*. Chicago: University of Chicago Press, 1980.

Maeda, John. *The Laws of Simplicity*. Boston: MIT Press, 2006.

Morville, Peter. *Ambient Findability: What We Find Changes Who We Become*. Sebastopol, CA: O'Reilly Media, 2005.

Musashi, Myamoto. *The Book of Five Rings.* New York: Bantam Books, 1982.

Norman, Don. *Design of Everyday Things.* Revised and expanded ed. New York: Basic Books, 2013.

———. *Living with Complexity.* Boston: MIT Press, 2010.

Petroski, Henry. *The Evolution of Useful Things: How Everyday Artifacts—From Forks and Pins to Paper Clips and Zippers—Came to Be as They Are.* New York: Vintage, 1994.

———. *Small Things Considered: Why There Is No Perfect Design.* New York: Vintage, 2004.

Pink, Daniel H. *A Whole New Mind: Why Right-Brainers Will Rule the Future.* New York: Riverhead, 2005.

Pye, David. *The Nature and Art of Workmanship.* New York: Cambridge University Press, 1968.

Reynolds, Garr. *Presentation Zen: Simple Ideas on Presentation Design and Delivery.* New York: New Rider, 2008.

Schön, Donald. *The Reflective Practitioner.* New York: Basic Books, 1984.

Semler, Ricardo. *The Seven-Day Weekend: Changing the Way Work Works.* New York: Portfolio, 2003.

Suzuki, Shunryu. *Zen Mind, Beginner's Mind.* Berkeley, CA: Weatherhill, 1970.

Cliff Crego's Poetry is available online at picture-poems.com/week4/complexity.html.

ABOUT THE AUTHOR

DAN WARD retired as a lieutenant colonel in the U.S. Air Force after spending more than two decades researching, developing, designing, testing, and fielding military equipment. He holds three engineering degrees and specializes in rapid, low-cost innovation. His assignments include the Air Force Research Lab, the National Geospatial-Intelligence Agency, the Air Force Institute of Technology, the Pentagon, and International Security Assistance Force Headquarters in Kabul, Afghanistan. Dan was awarded the Master Acquisition Badge and the Command Space Badge. In 2012, he received the Bronze Star Medal for his service in Afghanistan. He lives in Massachusetts with his wife and two daughters.